Oblivious Network Routing

Oblivious Network Routing

Algorithms and Applications

S. S. Iyengar and Kianoosh G. Boroojeni

The MIT Press
Cambridge, Massachusetts
London, England

MIT Press books may be purchased at special quantity discounts for business or sales promotional use. For information, please email special_sales@mitpress.mit.edu

This book was set in ITC Stone Serif Std 9/13pt by Toppan Best-set Premedia Limited, Hong Kong. Printed and bound in the United States of America.

Library of Congress Cataloging-in-Publication Data.

Oblivious network routing : algorithms and applications / S.S. Iyengar and Kianoosh G. Boroojeni.
    pages cm
Includes bibliographical Suggested Reading and index.
ISBN 978-0-262-02915-5 (hardcover : alk. paper) 1. Adaptive routing (Computer network management) I. Iyengar, S. S.
(Sundararaja S.) II. Boroojeni, Kianoosh G., 1989–
TK5105.54873.A33O25 2015
004.6—dc23
        2014034379

10 9 8 7 6 5 4 3 2 1

Contents

Preface

World-wide advances in transportation and telecommunication infrastructure have significantly changed human life over the last decades. As a result, we see an increasing exchange of national and international resources in today's world, which has turned our planet into a global village.[1] The process of international integration has substantially affected almost every aspect of human life, including economic activities, learning, communication, nutrition, and clothing, among others.

Our huge integrated world calls for an inexpensive, fast, and reliable way of transferring information, goods, and so on. The Internet, as a world-wide network, connects people around the world so that they can communicate with each other in a real-time fashion. In order to make the Internet faster, cheaper, and more reliable, there are a lot of challenges to deal with, such as the huge size of data flows and networks, diverse data traffic patterns locally and temporally, inappropriate infrastructure in some areas, political conflicts, and natural disasters that are unpredictable and inevitable. Similar challenges exist regarding other world-wide connecting networks, such as ground/air transportation and energy distribution networks. Consequently, there exists an increasingly important need for intelligent and *adaptable* routing of network flows.

The great impact of developing versatile solutions to the problem of routing network flows in unpredictable circumstances necessitated a large body of research work over the last few years. These attempts have evolved a rich literature in the context of "oblivious network design." This approach of routing the network flows usually leads to a routing scheme in the form of a spanning tree or a set of trees on the graph representation of the network.

---

1. Global Village is a term closely associated with Marshall McLuhan, popularized in his books *The Gutenberg Galaxy: The Making of Typographic Man* (1962) and *Understanding Media* (1964).

The common characteristics of the versatile routing schemes include being relatively flexible to the obliviousness of the environment in which the commodities are flowing and making the traffic flows distributed over the network; and preventing flow congestion in some specific nodes or edges. Additionally, these types of routing schemes provide a low-cost flow routing over the long term, even if a wide range of unpredictable events occurs in the network, such as flow bursts derived from a specific node or the failure of some node in forwarding the flow through the network. In fact, versatile routing schemes best fit to those networks that we have little or no knowledge regarding their current and future states. In this research monograph, we present some versatile routing schemes appropriate for proposing solutions to the routing problems that are oblivious in the context of the source, target, and value of the network flows.

In chapter 1, we provide an introduction to the network design. This introductory chapter has been designed to clarify the importance and position of the oblivious routing problems in the context of the network design as its containing field of research.

The "Mathematical Foundation" part of the monograph has been structured using two chapters: "Hierarchical Routing Tools and Data Structures" and "Routing Schemes in Oblivious Network Design." In the earlier chapter, we have made a fundamental discussion on the role of the *linked hierarchical data structures* in providing the mathematical tools needed to construct rigorous versatile routing schemes. This discussion is divided into two branches: the top-down hierarchical tools and the bottom-up tools.

Additionally, in chapter 3, we have applied the hierarchical routing tools in the process of constructing the versatile routing schemes. This has been done in a general way such that it can be used as an inspiration for designing the future schemes. Moreover, every claim regarding the routing schemes has been mathematically shown using a rigorous proof.

In the "Applications" part of the monograph, we have chosen two important applications of the versatile routing schemes. In chapter 4, a secure versatile model for content-centric networks has been evolved. This model helps the concept of versatility in the routing schemes to provide a congestion-free protocol for the content-centric networks that are assumed to play a key role in the future Internet.

As the second application of the versatile routing schemes, an interesting cost-optimization problem has been introduced in the context of the smart grids. We have proposed a novel approach for the use of the green power resources in a residential electricity system in a reliable and efficient way. This approach, which uses the versatile routing schemes, fairly balances the electric load over the system and at the same time, proposes a low-cost energy flow through the network over the long run.

Features

There are several unique aspects of our book that address the oblivious network routing problems:

1. The book specifies the importance and position of the oblivious network routing problems in the context of the network design as its containing field of research. To do this, numerous concepts in this area are defined precisely using the mathematical tools.
2. The book provides the basics and mathematical foundation needed to analyze and address the oblivious network routing problems. More specifically, it introduces some advanced data structures and tools (mostly in graph theory) that will be deployed for analysis and design of versatile routing schemes for oblivious routing problems.
3. The book presents some algorithms to generate some versatile routing schemes for oblivious routing problems by deploying the aforementioned advanced data structures. Additionally, the mathematical analysis of the scheme will be presented in detail.
4. To the best of our knowledge, this monograph applies the aforementioned routing techniques to the content-centric networks (which may be the future architecture of the Internet) for the first time.
5. To the best of our knowledge, this monograph applies the aforementioned routing techniques to the context of energy distribution in smart grids for the first time.

Acknowledgments

This work has evolved from our research on oblivious network design in distributed environments around 2006 from LSU with continuous body work which Dr. Iyengar began with his former student and colleague, Srivathsan Srinivasagopalan and Costas Busch, as well as others. It grew out of a research project supported by the Office of Naval Research, Defense Advanced Research Project, National Science Foundation, and other agencies. Financial support received from these agencies is gratefully acknowledged. The authors would like to thank Ivana Rodriguez for her helpful comments in improving the readability and presentation of this monograph. Professor Iyengar wishes to thank many students including Srivathsan Srinivasagopalan and Costas Busch for their contributions. He would also like to thank Richard Brooks and Vasanth Iyer for their careful review of the manuscript. Kianoosh G. Boroojeni would like to express his sincere gratitude to his wonderful family (Fariba, Mehrab, and Kaveh) and colleagues Ramin Ghadami, Roozbeh Nikkhah Moshaie, Farhad Hemmati, and Mani Shafa'atdoost for their continuous inspiration and support.

This project was partially supported by NSF Grant (CNS 1158701, CNS-0963793).

Chapter 1   Introduction to Network Design

A network is a common concept that is used to describe a group of interconnected things. These connections allow for individuals to communicate and cooperate by passing information flows and content flows through the various paths within a given network. In fact, we are all surrounded by networks. From scientific discoveries to natural phenomena, the concept of a network exists in biological, physical, and chemical systems, as well as a variety of distributions systems for transportation, telecommunication, and energy. For instance, a computer network such as the Internet has a vast number of cyber devices that transfer flows of data between computer hosts all around the world.

Figure 1.1 exemplifies yet another familiar network representation, an outline of U.S. highways throughout Pennsylvania in the late 1920s. In this ground transportation network, cities are interconnected by roads through which vehicles pass as network flows. Building an efficient ground transportation network and optimizing flow routes is critical and leads to a substantial savings in time and flow costs, especially over the long run. Similar concerns exist in telecommunication networks as optimized routing algorithms in Internet routers reduce electrical costs and create a considerable increase in network bandwidth.

In this chapter, we will examine a network as a *tool for flowing something from one place to another via the network connections in a step by step manner*. Each network flow will have the following properties: source point (initial place), target point (final place), flow path (the sequence of network connections that connect the end points), flow amount (the magnitude of whatever is flowing), and flow cost (the incurred cost of flowing). Additionally, we will focus on mathematically formulating the problem of routing flows through a network and the problems associated with cost minimization. To do this, we will develop a general framework to precisely define flow routing cost in a network. Moreover, a formal

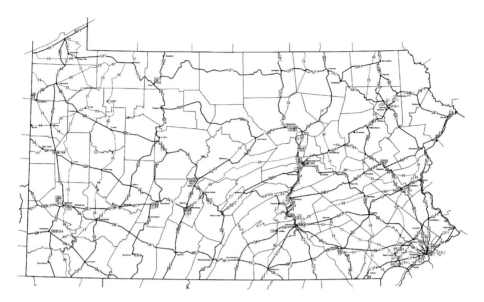

Figure 1.1 Pennsylvania's US Highways in 1928. Map by Timothy Reichard.

definition of obliviousness[1] in the context of network flows will also be presented.

1.1 Single-Source Network Routing Problem

In this section, the single-source routing problem in an undirected graph, which is a common representation of some networks, will be defined and illustrated through a series of examples. In addition, the flow cost of its solution will be discussed in detail. We will also address an extended version of this problem with multi-commodity flow.

1.1.1 Preliminary Definitions

Before starting our discussion on the network routing problem, we need to examine a few preliminary definitions.

Undirected graph G is defined as the ordered pair (V, E) such that V and E are the set of vertices and edges in G, respectively. Moreover, for every edge e in E, $e = \{u, v\}$ where u and v are two *distinct* vertices which are members of V. Edge e is called the connecting edge of vertices u and v if $e = \{u, v\}$.

1. The state of not being certain about what is happening.

For some undirected graph, a *simple path* is inductively defined in the following form:

Definition 1.1 In the undirected graph $G = (V, E)$, set $p \subseteq E$ is called a *(simple) path* from $s \in V$ to $t \in V$ if:

$$p = \begin{cases} \varnothing & \text{if } s = t \\ \{\{s, v\}\} \cup p' & \text{otherwise} \end{cases}$$

where $\{s, v\} \in E$ and p' is a path from v to t such that:

$$\forall e \in p' : s \notin e \tag{1.1}$$

Undirected graph G is *connected* if for every pair of vertices $u, v \in V$, there is a path from u to v.

In the above definition, if we ignore the constraint mentioned in equation (1.1), the obtained set p will be a *walk* of *no repetitive edge*. Additionally, if p denotes a path from s to t, it will also be a path from t to s; therefore, we call p a path *between s and t*. Furthermore, the path between a pair of vertices in G is not necessarily unique.

Now, consider the problem of finding paths from a given *source* vertex to a number of given *target* vertices in a connected undirected graph. There are many real-world examples that can be represented by such a problem. Here, we address a simplified example in the context of ground transportation.

Ground Transportation Example Assume that a company produces its products in a single factory and sells them in a number of retail outlets. The factory is located in city s, and there are a number of outlets scattered over different cities (there is at most one outlet per city). After making the products in the factory, they have to be distributed among the outlets. To facilitate the distribution, the products are first packed into boxes of the same size in the factory. Then, the boxes are taken from the factory to the outlets using ground transportation. Figure 1.2 schematically shows the ground transportation network through which the products are passed. This figure shows different cities including those in which the factory and outlets are located. It also specifies the available roads that can be used to take the products from one city to another.

Assume that each retail outlet needs a single box of a particular product to be taken from the factory. The goal is to satisfy the demand of each outlet by taking its needed box from the factory in city s to its designated target. This problem is an example of the *single-source network routing problem*, which we now formally define.

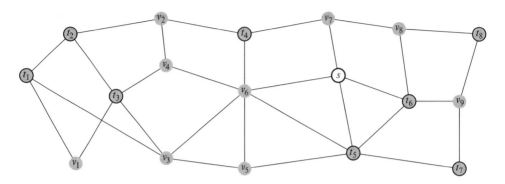

Figure 1.2 The schematic view of the transportation network. Circle s represents the city in which the factory located. Circles t_1, t_2, ..., t_8 are symbols of eight cities associated with eight retail outlets. The other circles represent the other cities that may participate in the paths from the factory to the outlets. The lines between the circles show the available roads between the cities.

Definition 1.2 Consider connected undirected graph $G = (V, E)$. Let $T = \{t_1, t_2, ..., t_k\}$ denote the set of targets where t_1, t_2, ..., t_k are k distinct *target* vertices that belong to V. Moreover, assume that $s \in V$ represents the *source* vertex. By definition, the *single-source network routing problem* is the problem of finding function $p:T \mapsto 2^E$ that maps target t_i to a simple path from s to t_i in graph G ($\forall i = 1, ..., k$). Function p and set $\{p(t):t \in T\}$ are called the solution function and the (feasible) solution of the problem, respectively.

Regarding figure 1.2, in the ground transportation example of the single-source network routing problem, V is the set of all cities, E is the set of roads between cities, and $T = \{t_1, t_2, ..., t_8\}$ is the set containing the cities in which the retail outlets are located. Moreover, the factory is located in the source city s. The solution is the set of paths $\{p(t_i):i = 1, ..., 8\}$ such that p is the solution function. As a result, for every $i = 1, ..., 8$, a product box is routed through the path $p(t_i)$ to reach the outlet t_i. Figure 1.3 represents one feasible solution by specifying all of the eight paths (note that the feasible solution of a single-source network routing problem is not unique). Note from figure 1.3 that each road belongs to a number of solution paths. For example, the road $\{t_6, v_9\}$ belongs to the paths $p(t_7)$ and $p(t_8)$. On the other hand, no solution path uses the road $\{t_5, t_7\}$. The number of paths to which a road belongs is called its *traffic flow*.

Definition 1.3 Consider the single-source network routing problem of target set T in graph $G = (V, E)$. Moreover, let p denote a solution function

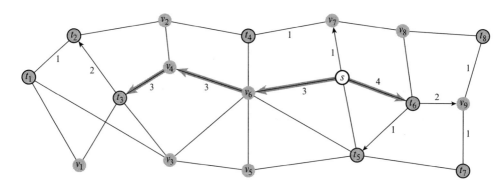

Figure 1.3 A solution of the single-source network routing problem in the ground transportation example. The highlighted lines show the paths through which the boxes are routed from the source to the designated targets. The number written on each highlighted line shows its traffic flow.

of the problem. Function $f_p: E \mapsto \mathbb{Z}_{\geq 0}$ will be the *flow function* associated with p if it is the case that:

$$f_p(e) = |\{p(t): t \in T, e \in p(t)\}| \quad \forall e \in E$$

The value $f_p(e)$ is called the *traffic flow of edge e* in solution $P = \{p(t): t \in T\}$.

Regarding Definition 1.3, the number of paths that use edge e in solution P specifies the traffic flow of e. For example, in the ground transportation problem, since each path is used to take one box, the value $f_p(e)$ is the number of boxes that are routed through the edge e in solution P. In figure 1.3, each road has been labeled with its traffic flow computed by the flow function f_p where p is the solution function corresponding to the shown solution. Note that there is no flow in the edges with no label (i.e., their traffic flow value is zero).

1.1.2 Edge Routing Cost

Previously, the single-source network routing problem was defined, and the ground transportation problem was described to illustrate the problem and its solution. Later in this section, the routing cost in this type of problem will be introduced and addressed in detail.

Let's consider the ground transportation example again. As mentioned earlier, in this problem, there are a number of retail outlets, say k, such that each one needs a box of product supplied by the single factory. The goal is to take all of the boxes to their appropriate demanding outlets. To do this, the company uses a number of trucks. Each truck is used to carry at most t

boxes from city x to its directly connected city y using the road $\{x, y\}$. Each box may be passed through different cities and carried by different trucks before reaching its demanding outlet. We will refer to the cost of using trucks to carry demands as the *routing cost*.

By modeling the problem as a single-source network routing problem, we have to specify k paths from the factory to each of the outlets. Assuming that p is a solution function of the problem, the total routing cost incurred by p can be obtained by first computing the cost incurred in each road, and then finding the total solution cost as an aggregation of all the computed costs.

First, our focus is on computing the cost incurred in each road. In this problem, moving the trucks containing some boxes of the factory products through the roads incurs some cost. Assuming the solution function p, each road e belongs to $f_p(e)$ paths in the solution (see Definition 1.3). As a result, there are $f_p(e)$ boxes that have to be passed through e. Since each truck can carry at most t boxes, the number of trucks needed to take $f_p(e)$ boxes through the road e is $\lceil f_p(e)/t \rceil$.

Assume that passing a truck through each road incurs some cost directly proportional to the road length. Subsequently, passing trucks through different roads incurs different cost amounts. Let c_e denote the cost of passing one truck through road e. Understanding that $\lceil f_p(e)/t \rceil$ trucks are passing through e, the total cost incurred in e will be $\lceil c_e \cdot f_p(e)/t \rceil$. In addition, let a denote the cost of loading each box into a truck at the beginning of each road and unloading it from the truck at the end of it. As a result, the total cost of routing demand flows through road e can be written as a function of its traffic flow.

$$\text{cost}_e(x) = \frac{x}{t} \cdot c_e + a \cdot x \quad \text{where } x = f_p(e) \tag{1.2}$$

In this equation, the function cost_e is called the *cost function* of road e (different roads have different cost functions). Figure 1.4 shows the scatter plot of the routing cost incurred in road e versus its traffic flow for the specific configuration of our ground transportation example. In general, the routing cost incurred in each edge is defined in the following way (for the single-source routing problem).

Definition 1.4 Consider the single-source network routing problem of graph $G = (V, E)$. Assuming that edge e has traffic flow f in some feasible solution, the value of $\text{cost}_e(f)$ is defined as the routing cost incurred by the solution in edge e where function $\text{cost}_e \colon \mathbb{Z}_{\geq 0} \mapsto \mathbb{R}_{\geq 0}$ is called the *edge routing cost function* for every edge $e \in E$.

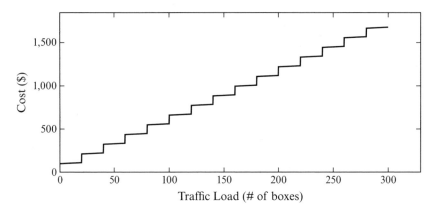

Figure 1.4 The plot of the routing cost function of road e in the ground transportation example for truck capacity of $t = 20$, loading and unloading cost of $a = \$0.60$, and truck passing cost of $c_e = \$100$ through road e.

From Definition 1.4, the routing cost value of edge e is a function of both the edge and its traffic flow. In many real-world examples, the edge routing cost function is *multiplicatively separable*, i.e., it can be written as the product of two single input functions. For example, in the ground transportation problem, assuming that the value of a in equation (1.2) is negligible (in comparison with c_e/t), $\mathrm{cost}_e(x)$ will be the product of $\lceil x/t \rceil$ (which is a function of traffic flow) and c_e (which only depends on the road properties).

Definition 1.5 The edge routing cost function cost_e of the single-source network routing problem is said to be (multiplicatively) separable if:

$$\mathrm{cost}_e(x) = \mathrm{rrc}(x) \cdot w_e \qquad (1.3)$$

In this equation, w_e is called the *weight* of edge e for every $e \in E$, and the function $\mathrm{rrc} \colon \mathbb{Z}_{\geq 0} \mapsto \mathbb{R}_{\geq 0}$ is called the *relative edge routing cost function* or simply the *relative routing cost function*.

In equation (1.3), function rrc determines how the routing cost of an edge is related to its traffic flow value. This relation is identical for every edge in the problems with separable edge routing cost function. The *absolute* routing cost of each edge in these problems is the weighted (scaled) value of its relative routing cost.

In the ground transportation problem with a separable edge routing cost function, the edge weight of each road is c_e, which denotes the cost of passing a truck through that road. Moreover, the relative routing cost

Figure 1.5 The schematic view of a link and two routers. Data packets enter router *a* with the average rate λ and exit with the average rate μ.

function in this problem is ⌈*x*/*t*⌉, which is a concave function.[2] Later in this chapter, we will see that the rrc function is the critical parameter in routing cost optimization problems in such a way that its properties may significantly affect the complexity of problem solving algorithms. In the following example, we will see that the rrc function is convex.[3]

Assume that there is a computer network that consists of a set of routers and links connecting each pair of routers. Consider graph $G = (V, E)$ as the model of the computer network where V represents the set of identical routers and E models the set of identical links between them. Let e denote a link between routers a and b. Figure 1.5 schematically shows the link and the routers and the data packet flow direction from a to b. Although there is a two-way data packet flow in the link, our focus is on the direction shown in figure 1.5. In this figure, the rate of exiting packets from the waiting queue of router a is assumed to be μ packets per second, which is referred to as the link *transmission speed*. Moreover, the process of packets arriving at the waiting queue of router a is assumed to be a Poisson process[4] of intensity λ packets per second. Consequently, the average traffic flow of packets in the specific time interval $[t_0, t_0 + n]$ is equal to the product of intensity λ and the length of the interval (n). As a result, considering $\bar{f}_{[t,t+n]}$ to be the average traffic flow in link e, we obtain the following equation:

$$\lambda = \frac{\bar{f}_{[t,t+n]}}{n} \tag{1.4}$$

In this example, the value of the edge routing cost function cost$_e$ for traffic flow of f packets into edge e (which denotes the link shown in figure 1.5) is equal to f times the cost of routing one packet through e. Assume

2. The real-valued function f on the interval $[x, y] \in \mathbb{R}$ is concave if and only if $\forall t \in [0, 1]$: $f(tx + (1 - t)y) \geq tf(x) + (1 - t)f(y)$.
3. The real-valued function f on the interval $[x, y] \in \mathbb{R}$ is convex if and only if $\forall t \in [0, 1]$: $f(tx + (1 - t)y) \geq tf(x) + (1 - t)f(y)$.
4. Poisson process of intensity λ is a stochastic continuous-time process N in that its value in arbitrary time t is a stochastic, discrete, Poisson distributed variable such that $\forall t > 0$: $N_t \sim \text{Pois}(\lambda t)$.

that the cost of taking a packet from router a to router b is directly proportional to the length of time interval from the moment of entering the packet into the waiting queue of router a to the moment when it reaches the same point of router b (i.e., the entrance of the waiting queue in router b). This interval consists of three parts: waiting time in the queue of router a, transmission delay,[5] and propagation delay.[6] Assuming that propagation delay is negligible compared with the other two delays, the cost of routing each packet through edge e and also the function $cost_e$ can be written in the following form:

$$\text{cost of routing each packet through } e =$$
$$c \cdot (\text{transmission delay} + \text{queuing delay})$$

or, equivalently:

$$\text{cost}_e(f) = f \cdot c \cdot (\text{transmission delay} + \text{queuing delay}) \tag{1.5}$$

where c is the coefficient of the proportionality relation between the routing cost incurred and the routing time spent in the process of routing packets through a link.

To compute the queuing delay in router a, we use a preliminary theorem of queue theory. Since packets enter the queue of router a in a Poisson process of intensity λ packets per second, the waiting time of packets in such a queue follows the exponential distribution of rate $(\mu - \lambda)$ packets per second. Subsequently, the average queue waiting time for each packet is $1/(\mu - \lambda)$. Moreover, since the link transmission speed is μ, the average transmission delay for each packet is $1/\mu$. Using the mentioned average delays, we can rewrite equation (1.5) in the following form:

$$\text{cost}_e(f) = f \cdot c \cdot \left(\frac{1}{\mu} + \frac{1}{\mu - \lambda} \right) \tag{1.6}$$

Replacing λ in equation (1.6) with its equivalent expression from equation (1.4) produces the following formula (again, we use the average value of traffic flow and f interchangeably).

$$\text{cost}_e(f) = f \cdot c \cdot \left(\frac{1}{\mu} + \frac{n}{n\mu - f} \right) \tag{1.7}$$

Equation (1.7) represents the average routing cost of edge e as a function of its average traffic flow f over an interval of n seconds. In this equation,

5. The amount of time it takes for a router to transmit all the bits of a packet over a link connected to it.
6. The amount of time it takes for all the bits of a packet to move from one end of a link to the other.

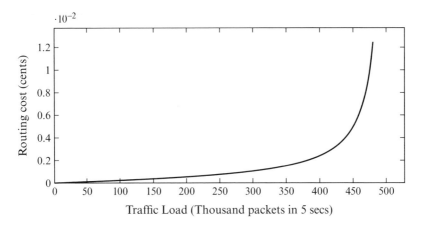

Figure 1.6 The plot of the relative routing cost function of the computer network problem for $n = 5$ sec, $c = \$10^6$, and $\mu = 10^5$ packet/sec.

because of the assumption that all the routers and the links are identical, the edge routing cost function is the same for every link e in the computer network. In fact, parameters c and μ are independent of e. Consequently, it is safe to assume that the edge weight $w_e = 1$ for all the links. This implies that rrc $\equiv$ cost$_e$ in this problem. Moreover, concerning equation (1.7), the rrc function of the computer network problem is convex (despite the ground transportation problem). Figure 1.6 plots the rrc function of the computer network problem.

1.1.3   Network Routing Cost

Previously, edge routing cost was defined by two functions, cost$_e$ and rrc, in the single-source network routing problem. In this section, we formally define the *network routing cost* which specifies how costly a solution of some network routing problem is. In spite of the edge routing cost, which is defined for every edge of the graph, the network routing cost computes the aggregated cost incurred over all of the edges. In fact, function cost$_e$ determines the relation between the routing cost incurred in e and its traffic flow based on the characteristics of e rather than considering the whole graph's topology. On the other hand, the network routing cost of a solution depends on the routing cost of all of the edges in the graph. We now give a formal definition of the network routing cost.

Definition 1.6   Consider the single-source network routing problem of graph $G = (V, E)$ and solution function p. By definition, function $nrc_\pi: (\mathbb{R}_{\geq 0})^{|E|} \mapsto \mathbb{R}_{\geq 0}$ is called the *network routing cost function* such that $\pi:\{1, 2,$

..., $|E|\} \mapsto E$ denotes an arbitrary permutation[7] of E members. Assuming that the routing cost of edge e corresponding to solution function p is represented by $C_p(e)$ (for every $e \in E$), the associated *network routing cost* C_p is then defined by the following equation:

$$C_p = \mathrm{nrc}_\pi\left(C_p(\pi_1), C_p(\pi_2), \ldots, C_p(\pi_{|E|})\right)$$

Now, consider the ground transportation example again. In this example, since we are concerned with the total transportation cost, we define the network routing cost of solution function p as the summation of the routing costs in all the roads. As a result, concerning Definition 1.6, the value of C_p in this example is obtained by the following equation:

$$C_p = \sum_{i=1}^{|E|} C_p(\pi_i) = \sum_{e \in E} C_p(e) \tag{1.8}$$

where $C_p(e)$ is the routing cost of edge (road) e incurred by p. Regarding Definition 1.4, the routing cost of edge e is the output of function cost_e, which takes the traffic flow of edge e as the input. Since the traffic flow of e is shown by $f_p(e)$, we obtain that:

$$C_p(e) = \mathrm{cost}_e\left(f_p(e)\right)$$

By replacing the equivalent expression of $C_p(e)$ in equation (1.8), we arrive at the following equation:

$$C_p = \sum_{e \in E} \mathrm{cost}_e\left(f_p(e)\right)$$

Moreover, in the computer network example described previously, the network routing cost is assumed to be the total cost incurred over all of the routers and their connecting links. Since the incurred cost in every router and its connected link was previously defined as the edge routing cost, we define the network routing cost as the summation of all the edge routing costs. In consequence, equation (1.8) also works for the computer network example. In fact, in both examples, function nrc_π outputs the simple addition of its input arguments:

$$\mathrm{nrc}_\pi\left(C_p(\pi_1), C_p(\pi_2), \ldots, C_p(\pi_{|E|})\right) = \sum_{i=1}^{|E|} C_p(\pi_i) \tag{1.9}$$

In general, the network routing cost function presented in equation (1.9) can be used in many examples of the network routing problem. Another common nrc_π function is the max function, which is used when the concern is about the *flow congestion* in network edges (connections).

$$C_p = \mathrm{nrc}_\pi\left(C_p(\pi_1), C_{p(\pi_2)}, \ldots, C_p(\pi_{|E|})\right) = \max_{i=1,2,\ldots,|E|} C_p(\pi_i)$$

---

7. Function $\pi:\{1, 2, \ldots, n\} \mapsto A$ is a permutation of A members if $|A| = n$, and π is a one-to-one function. Notation π_i is commonly used to represent the output of function π corresponding to value i.

1.2   General Network Routing Problem

Earlier in this chapter, we addressed the single-source network routing problem and the cost of its solutions. In addition, some examples were made to illustrate the discussion. This section is about a more general version of network routing problems. Despite the aforementioned problem, every graph vertex may send flow through the network in the general form. Moreover, each flow has some value that affects the traffic flow amount of edges through which the flow is passing. Here, we present some basic definitions.

Definition 1.7   Let $G = (V, E)$ denote an undirected graph.

- The triple $(s, t, d) \in (V \times V \times \mathbb{R}^+)$ is called a commodity in G. The vertices s and t are the source and target of the commodity, respectively. Moreover, d is called the commodity value.
- Consider $\bar{K} = (K_1, K_2, \ldots, K_k)$ to be a sequence of commodities in graph G such that for every $i, j = 1, 2, \ldots, k$, if $K_i = (s, t, d)$ and $K_j = (s', t', d')$, the following condition holds:

$$(s = s' \wedge t = t') \rightarrow i = j$$

By definition, the *general network routing problem* or simply *general routing problem* is the problem of finding a sequence of paths like $\bar{p} = (p_1, p_2, \ldots, p_k)$ such that for every $i = 1, \ldots, k$, p_i represents a simple path from $\Pi_1(K_i)$ to $\Pi_2(K_i)$ in graph G.[8] Sequence $\bar{p}$ is called an *integral solution* of the problem.

Similar to the single-source network routing problem, here we define the traffic flow of edges in the general version.

Definition 1.8   Assuming that $\bar{p}$ is an integral solution for the general routing problem of graph $G = (V, E)$, the function $f_{\bar{p}}^{(i)}: E \mapsto \mathbb{R}_{\geq 0}$ is *the flow function* of the ith commodity in solution $\bar{p}$ if:

$$f_{\bar{p}}^{(i)}(e) = \begin{cases} d_i & \text{if } e \in p_i \\ 0 & \text{otherwise} \end{cases} \quad \forall e \in E, \forall i = 1, 2, \ldots, k \qquad (1.10)$$

The value of $f_{\bar{p}}^{(i)}(e)$ is called the *traffic flow* of commodity i through edge e.

Note that from Definition 1.8, if there are k commodities, we define k flow functions for every solution $\bar{p}$: $f_{\bar{p}}^{(1)}, f_{\bar{p}}^{(2)}, \ldots, f_{\bar{p}}^{(k)}$; however, in the single-source version of the problem, we only defined one flow function for every solution.

---

8. Assuming $x = (x_1, x_2, \ldots, x_n)$ as an ordered n-tuple, for every $k = 1, 2, \ldots, n$, if projection operation Π_k gets x as the input, the output will be $\Pi_k(x) = x_k$.

To define the solution cost, similar to the single-source version, we need to first define the routing cost incurred in every edge, and then compute the network routing cost based on the routing costs of all the edges.

Definition 1.9   Consider the general routing problem of graph $G = (V, E)$ and solution $\bar{p}$. Assuming that there are k commodities in the problem, function $\text{cost}_e \colon (\mathbb{R}_{\geq 0})^k \mapsto \mathbb{R}_{\geq 0}$ is called the *edge routing cost function* for every edge $e \in E$. Moreover, if f_i denotes the traffic flow of commodity i through edge e in $\bar{p}$ (for every $i = 1, 2, \ldots, k$), the routing cost of e incurred by solution $\bar{p}$ is then defined as $\text{cost}_e(f_1, f_2, \ldots, f_k)$.

Note that in the above definition, the traffic flow f_i is obtained by the following equation:

$$f_i = f_{\bar{p}}^{(i)}(e) \quad \forall i = 1, 2, \ldots, k$$

The network routing cost function nrc_π for the general routing problem is defined in a similar fashion to what was mentioned in Definition 1.6 for the single-source version of the problem. As an example of the general routing problem, consider a computer network that connects some hosts together. To provide this connection, it uses some routers and links. Each link connects either a host to a router or a pair of routers to each other. Figure 1.7 shows the graph representation of this network.

As figure 1.7 shows, the routers and hosts are represented as the graph vertices, and the links between them are modeled as the graph edges.

Assume that the hosts are communicating with each other using k different protocols $r_1, r_2, \ldots, r_k$. Also, assume that k connections $c_1, c_2, \ldots, c_k$ have already been established between hosts such that connection c_i connects the hosts s_i and t_i with protocol r_i for every $i = 1, 2, \ldots, k$. Hosts s_i and t_i are

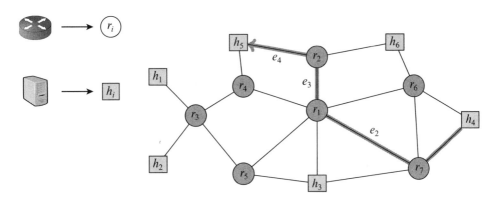

Figure 1.7 Computer network example: In this figure, a path of data flow in a connection from host h_4 to host h_5 has been highlighted. The path is $\{e_1, e_2, e_3, e_4\}$.

called the source and target hosts of connection c_i, respectively. Note that each host can simultaneously be the source or target of more than one connection. Moreover, statistical analysis shows that protocol r_i sends d_i bits (on average) from the connection source to its target. For simplicity and without loss of generality, we assume that every established communication is one-way (our proposed formulation is extendable to two-way communication). In addition, during a connection session, data is sent through a single simple path specified at the connection start-up. The problem of routing (finding the paths for) the data flow of the connection sessions can be represented by a general routing problem.

To formulate this problem, consider graph $G = (V, E)$ such that V is the set of cyber nodes (i.e., routers and hosts) and E denotes the set of all the links connecting the nodes. Moreover, let $\bar{K} = (K_1, K_2, \dots, K_k)$ denote the sequence of commodities such that K_i specifies the triple (s_i, t_i, d_i) for every $i = 1, 2, \dots, k$ ($s_i, t_i,$ and d_i are characteristics of the ith connection c_i). Assume that solution $\bar{p} = (p_1, p_2, \dots, p_k)$ specifies k paths in the network such that p_i is the set of links connecting the pair of hosts s_i and t_i. Since each link of the network may participate in more than one connection, we present data flow through the link e using k values $f_{\bar{p}}^{(1)}(e), f_{\bar{p}}^{(2)}(e), \dots, f_{\bar{p}}^{(k)}(e)$ where $f_{\bar{p}}^{(i)}(e)$ represents the average number of bits sent by connection c_i (which uses protocol r_i) through link e in one second.

Sending data through the network incurs some cost. Assume that there are two costly events during the data sending process:

1. The incoming data to a router waits in its waiting queue.
2. The router transmits data through the appropriate link connected to it.

Note that for simplicity, we ignore the cost of propagating data through the links and processing data in routers. Assuming that link e has transmission speed of μ_e bits per second and the process of arriving bits at each router is a Poisson process, the cost of data transmission and queue waiting delay of each router incurred in a one second period length is obtained by an equation similar to equation (1.7). We just need to replace μ by μ_e, and n by one in equation (1.7) to obtain the cost incurred in link e (note that in equation (1.7), the measurement unit of data amount was a data packet; however, in this example, it is a bit):

$$\text{The cost incurred in edge } e = f \cdot c \cdot \left(\frac{1}{\mu_e} + \frac{1}{\mu_e - f} \right) \tag{1.11}$$

In equation (1.11), c denotes the proportionality coefficient of the linear relation, which is assumed to exist between the routing time spent and the routing cost incurred in link e. In addition, in this equation, f denotes the average of total number of bits sent by all the connections through link e in a one second length period of time; i.e.:

$$f = \sum_{i=1}^{k} f_{\bar{p}}^{(i)}(e) \tag{1.12}$$

At this point, we take the cost of processing data into consideration. Assume that it costs a_i units for any router to process one data bit of protocol r_i. Subsequently, processing $f_{\bar{p}}^{(i)}(e)$ bits of the ith connection data incurs the amount of $a_i \cdot f_{\bar{p}}^{(i)}(e)$ units for the router, which is transmitting data into link e. As a result, regarding equations (1.11) and (1.12), the edge routing cost function of the problem can be written in the following form:

$$\text{cost}_e(f_1, f_2, \dots, f_k) = \sum_{i=1}^{k} a_i f_i + \left(c \sum_{i=1}^{k} f_i \right) \left(\frac{1}{\mu_e} + \frac{1}{\mu_e - \sum_{i=1}^{k} f_i} \right) \tag{1.13}$$

Note that equation (1.13) is only true for those links that are not connected to the source hosts of connections: $s_1, s_2, \dots, s_k$. For the link connected to source host s_i (for some $i = 1, 2, \dots, k$), since there is no router at the source end of the link, the queue waiting cost and the router processing cost is zero, and the only incurred cost is the transmission cost, which equals (transmission speed in link e is μ_e):

$$\text{cost}_e(f_1, f_2, \dots, f_k) = \frac{c \sum_{i=1}^{k} f_i}{\mu_e} \tag{1.14}$$

As an example, consider the network represented in figure 1.7. The function cost_{e_1} is defined by equation (1.14); however, the routing costs of links e_2, e_3, and e_4 are obtained by equation (1.13).

The last step in modeling our example by the general routing problem is to define the network routing cost function nrc_π. If we are concerned with the data congestion in the connecting links, we have to define nrc_π as the max function. However, if we are concerned with the total routing cost, nrc_π must be the summation of all of the edge routing costs.

Routing Cost Optimization   In a general routing problem of graph $G = (V, E)$, assume that:

- $g_e(f)$ is the edge routing cost function for every edge $e \in E$ and flow $f \geq 0$.
- $h_\pi(x_1, x_2, \dots, x_{|E|})$ denotes the network routing cost function for $x_i \geq 0$ and $i = 1, 2, \dots, |E|$.
- $\bar{K} = (K_1, K_2, \dots, K_k)$ is the sequence of commodities.

By definition, the triple $\mathcal{E} = \left(g_e, h_\pi, \bar{K} \right)$ is called the *routing cost environment* of the problem. The routing cost environment specifies all the commodities in a general routing problem and also provides a tool for evaluating the routing costs of its solution. Here, we formulate the *routing cost optimization problem* using our routing cost environment.

Definition 1.10   Consider the general routing problem of graph $G = (V, E)$ and routing cost environment $\mathcal{E} = \left(g_e, h_\pi, \bar{K}\right)$. The problem of finding an integral solution that minimizes the network routing cost in environment $\mathcal{E}$ is called the *routing cost optimization problem*. We define the minimum routing cost $C_{\mathcal{E}}^*$ in environment $\mathcal{E}$ in the following form:

$$C_E^* = \min_{\bar{p} \in \mathcal{I}_{\bar{K}}} h_\pi \left(C_{\bar{p}}(\pi_1), C_{\bar{p}}(\pi_2), \dots, C_{\bar{p}}(\pi_{|E|}) \right)$$

In this equation, $\mathcal{I}_{\bar{K}}$ denotes the set of all the integral solutions of the problem with the sequence $\bar{K}$ of commodities. Furthermore, function $C_{\bar{p}}(e)$ outputs the routing cost of edge e in solution $\bar{p}$ (where $f_{\bar{p}}^{(i)}$ denotes the flow function of the ith commodity in the solution $\bar{p}$):

$$C_{\bar{p}}(e) = g_e \left(f_{\bar{p}}^{(1)}(e), f_{\bar{p}}^{(2)}(e), \dots, f_{\bar{p}}^{(k)}(e) \right) \quad \forall e \in E$$

In Definition 1.10, the integral solution(s) which has the minimum network routing cost in environment $\mathcal{E}$ is called *the optimized integral solution* of the problem and denoted by $\bar{p}_{\mathcal{E}}^*$; utilizing this terminology, we obtain the following equation:

$$C_{\mathcal{E}}^* = h_\pi \left(C_{\bar{p}_{\mathcal{E}}^*}(\pi_1), C_{\bar{p}_{\mathcal{E}}^*}(\pi_2), \dots, C_{\bar{p}_{\mathcal{E}}^*}(\pi_{|E|}) \right)$$

1.3   Oblivious Routing Cost Environment

In the previous sections, we introduced two varieties of network routing problems. Here, we define another type of problem with an oblivious routing cost environment. To illustrate the obliviousness of the routing cost environment, we first make an example in the context of cyber networks; then the definition of *oblivious routing problem* will be presented formally.

Consider the Internet as a world-wide network. At any moment, a large amount of data is being exchanged between computer devices scattered all around the world. This means that many data flows with different source and target hosts are passing through the Internet at any given moment. Moreover, the traffic flow amount at each point located in the Internet is permanently changing, which eventuates in the time-varying cost amount incurred by the data flows passing through it; for example, in a highly busy data connection, it takes a relatively high amount of time for a data packet to be transferred through the connection. In consequence, the Internet is a network that is being used in a wide range of routing cost environments, i.e., the network flows may have numerous sources, targets, and values, and also they incur varying amounts of cost while passing through the network over time. Now, consider the following constraint of the Internet:

It is a network through which the low-cost flow paths are too time-consuming to be computed online (upon request). This is essentially due to the large size of the network and the necessity to route very quickly.

A practical way of routing flows in the Internet (and the networks with similar constraints) is to route the flows in an offline manner. In other words, as the flow routing process cannot be done in a *real-time* fashion, we have to *precompute* the flow paths. Additionally, as the routing cost environment in the Internet is varying from time to time, it is necessary to route the flows under *every possible* routing cost environment. This type of routing flow through a large-scale network is highly adaptable to the obliviousness of the routing cost environment and the huge problem size. Consequently, routing the flows in this way is said to be *versatile*.

Consider the computer network example again. Assume that each pair of hosts can unexpectedly initiate a connection with one of the protocols r_1, r_2, ..., r_k and start communicating with each other via the connection. In this network, the number of connections, and also the source, target, and protocol type of each connection, is *non-deterministic*. In other words, the sequence of commodities $\bar{K}$ in the associated general routing problem is oblivious. Moreover, the routing cost function of each connection (cost_e), which depends on the number of connections of each protocol, is also oblivious. For example, if we have $2k$ connections c_1, c_2, ..., c_{2k} at any given moment, and we know that connections c_{2i-1} and c_{2i} use protocol r_i ($\forall i = 1$, ..., k), the following equation holds:

$$\text{cost}_e(f_1, f_2, \dots, f_{2k}) = \sum_{i=1}^{2k} \left(a_{\lceil \frac{i}{2} \rceil} \cdot f_i \right) + \left(c \sum_{i=1}^{2k} f_i \right) \cdot \left(\frac{1}{\mu_e} + \frac{1}{\mu_e - \sum_{i=1}^{2k} f_i} \right)$$

which is completely different from the cost function obtained in equations (1.13) and (1.14). Consequently, we have to deal with a set of routing cost optimization problems with different edge routing cost functions for each.

Definition 1.11 Let $P_{G,\mathcal{E}}$ denote a general routing problem of graph $G = (V, E)$ in routing cost environment $\mathcal{E}$. The following set of general routing problems is called an *oblivious routing problem* if $|\mathbb{E}| > 1$.

$$\{P_{G,\mathcal{E}}: \mathcal{E} \in \mathbb{E}\}$$

Set $\mathbb{E}$ is called the *set of possible routing cost environments* of the oblivious routing problem.

As you see in Definition 1.11, the total number of possible routing cost environments equal to $|\mathbb{E}|$ has to be more than one. In fact, if we only have one possible routing cost environment, the problem has no obliviousness and it turns into a general routing problem.

Definition 1.12 Consider an oblivious routing problem of graph $G = (V, E)$ and the set of possible routing cost environments $\mathbb{E}$. An *integral solution* of the oblivious routing problem is defined as the function:

$$S: \mathbb{E} \mapsto \bigcup_{E \in \mathbb{E}} \mathcal{I}_{\bar{K}_{\mathcal{E}}} \tag{1.15}$$

such that $S(\mathcal{E}) \in \mathcal{I}_{\bar{K}_{\mathcal{E}}}$. In relation (1.15), $\bar{K}_{\mathcal{E}}$ represents the third member of triple $\mathcal{E}$, i.e., $\Pi_3(\mathcal{E})$, and also $\mathcal{I}_{\bar{K}_{\mathcal{E}}}$ denotes the set containing all the integral solutions of the general routing problem with the sequence $\bar{K}_{\mathcal{E}}$ of commodities in graph G (for every $\mathcal{E} \in \mathbb{E}$).

In Definition 1.12, the integral solution $S(\mathcal{E})$ is the solution suggested by function S for problem $P_{G,\mathcal{E}}$. Additionally, note that instead of considering one path sequence as the solution of an oblivious routing problem, we have to specify a function that maps an integral solution to each of the possible routing cost environments. As there may be more than one solution for a general routing problem, the solution function of the oblivious routing problem is not necessarily unique as well. To compute the integral solution of a given oblivious routing problem, we introduce two different approaches.

Dynamic Approach In the dynamic approach, by considering one possible routing cost environment at a time, we compute the integral solution of the resulted general routing problem and do this process for all other possible routing cost environments in a repetitive manner. In other words, we take advantage of the fact that by restricting the routing cost environment to a special case, our oblivious routing problem will turn into a general routing problem. Subsequently, the integral solution function obtained by the dynamic approach is in the following form:

$$S_{\text{dynamic}}(\mathcal{E}) = \bar{p}_{\mathcal{E}} \quad \forall \mathcal{E} \in \mathbb{E}$$

where $\bar{p}_{\mathcal{E}}$ is an integral solution of problem $P_{G,\mathcal{E}}$. Moreover, in the case that we are concerned with optimizing the routing cost of the oblivious routing problem, we use the following equation:

$$S_{\text{dynamic}}(\mathcal{E}) = \bar{p}_{\mathcal{E}}^{*} \quad \forall \mathcal{E} \in \mathbb{E}$$

where $\bar{p}_{\mathcal{E}}^{*}$ is the optimized integral solution of problem $P_{G,\mathcal{E}}$.

Although this approach results in the best answer and minimizes the network routing cost for any given set of possible routing cost environments, it is computationally impractical in most real-world cases.

Versatile Approach   As a different option from the dynamic approach, the versatile approach is computationally much more efficient, but usually fails to minimize the network routing cost for every possible routing cost environment. This approach defines a *versatile routing scheme* to compute the integral solution function.

Definition 1.13   Consider an oblivious routing problem of graph $G = (V, E)$ and set $\mathbb{E}$ of possible routing cost environments. The *(integral) versatile routing scheme* is defined as the following function:

$$\mathbb{S}: V^2 \mapsto \text{the set of all the paths in } G$$

such that for every pair of vertices $u,v \in V$, $\mathbb{S}(u, v)$ specifies a simple path between u and v in G. The solution function associated with the versatile routing scheme $\mathbb{S}$ is represented as $S_{\mathbb{S}}$. Assuming $\left(\text{cost}_e, \text{nrc}_\pi, \overline{K}\right)$ as an arbitrary member of $\mathbb{E}$, it follows that, given $\overline{K} = (K_1, K_2, \dots, K_k)$:

$$S_{\mathbb{S}}\left(\text{cost}_e, \text{nrc}_\pi, \overline{K}\right) = (p_1, p_2, \dots, p_k)$$

such that $\mathbb{S}(\Pi_1(K_i), \Pi_1(K_i))$ for every $i = 1, 2, \dots, k$.

Consequently, in the versatile approach, we use the paths specified by function $\mathbb{S}$ to route every commodity of the possible routing cost environments. In chapter 3, we will define and analyze different versatile routing schemes.

In order to evaluate the cost efficiency of solution $S_{\mathbb{S}}$, we use the *competitive* approach, which compares the incurred cost of solution $S_{\mathbb{S}}$ with the cost incurred by the most efficient solution S_{dynamic} (which is obtained by the dynamic approach).

Consider Definition 1.13 again. For every $\mathcal{E} \in \mathbb{E}$, the competitiveness ratio of scheme $\mathbb{S}$ in the general routing problem $P_{G,\mathcal{E}}$ is defined in the following form:

$$\text{cr}(\mathcal{E}, \mathbb{S}) = \frac{C_{\mathcal{E}}(S_{\mathbb{S}}(\mathcal{E}))}{C_{\mathcal{E}}(\overline{p}_{\mathcal{E}}^*)}$$

such that $\overline{p}_{\mathcal{E}}^*$ is the optimized integral solution of problem $P_{G,\mathcal{E}}$ and function $C_{\mathcal{E}}$ outputs the network routing cost of its input solution of problem $P_{G,\mathcal{E}}$. Moreover, the competitiveness ratio of scheme $\mathbb{S}$ in the oblivious routing problem of set $\mathbb{E}$ of the possible routing cost environments is represented by $\text{CR}(\mathcal{E}, \mathbb{S})$ and defined by the following equation:

$$\text{CR}(\mathcal{E}, \mathbb{S}) = \max_{\mathcal{E} \in \mathbb{E}} \text{cr}(\mathcal{E}, \mathbb{S})$$

As a result, the value of the competitiveness ratio quantifies the cost efficiency of a versatile routing scheme in one or more routing cost environments.

1.4 Fractional versus Integral Routing

In Section 1.2, the general routing problem was formally defined. From Definition 1.7, in the general routing problem, we have to specify a simple path for each commodity so that it flows through the path to reach its target. Since we had to route k commodities in this problem, sequence $\bar{p}$ of k simple paths was specified as the integral solution. Moreover, remember that for the integral solution $\bar{p}$, $f_{\bar{p}}^{(i)}(e)$ denotes the flow value of commodity $K_i = (s_i, t_i, d_i)$ through edge e and is defined in equation (1.10). From this equation, $f_{\bar{p}}^{(i)}(e)$ can only get *two* values: zero and d_i. In the case that $f_{\bar{p}}^{(i)}(e)$, as edge e does not belong to p_i, e has not been used to route the ith commodity in solution $\bar{p}$. On the other hand, if $f_{\bar{p}}^{(i)}(e)$, since e is a member of p_i, the whole flow value is routed through e. Subsequently, our hidden assumption is that we do not break down a commodity into smaller parts before sending it through the network. Hence, in the *integral* solution of the general routing problem, the traffic flow of each commodity can only be routed through a *single* path.

On the other hand, there is another type of solution in which the flow of each commodity is not restricted to only one simple path. In other words, there may be multiple simple paths through which the commodity *fractions* flow from the source vertex to the target. As a result, the total value of each commodity (d_i) will be carried from the source point to the target. This solution of the general routing problem is known as the *fractional solution*.

For instance, consider the computer network example from Section 1.2 again. Assume that the network protocols allow the connection source to flow data toward the target through multiple paths rather than a single predefined path. In this case, the fractional solution is also acceptable for the routing problem.

In this section, we will formally present the fractional flow, the fractional solution of the general routing problem, and the fractional version of the routing cost optimization problem. The obliviousness in these problems will then be discussed.

Definition 1.14 Consider graph $G = (V, E)$ such that V and E represent the vertex and edge set of G, respectively. Fractional flow F of (total) value X, source vertex $u \in V$, and target vertex $v \in V$ are defined in the following form:

$$F = \{(p, x): x \in (0, X], p \text{ is a simple path in } G \text{ from } u \text{ to } v\}$$

such that:

1. $\big((p_1, x_1) \in F\big) \wedge \big((p_2, x_2) \in F\big) \wedge (p_1 = p_2) \rightarrow (x_1 = x_2)$.
2. $\sum_{(p,x)\in F} x = X$.

Henceforth, the fractional flow in graph G is a set of ordered pairs such that each one specifies a simple path (p) in G and a fraction that specifies the weight of path p. In Definition 1.14, condition 1 forces each member of flow F to represent different paths. Additionally, condition 2 guarantees that the total flow value (X) will eventually be carried by all the paths associated with F members. At this point, we can extend the definition of the general routing problem to the problem with fractional solution.

Definition 1.15   Consider graph $G = (V, E)$ and k distinct commodities $\bar{K} = (K_1, K_2, \dots, K_k)$ such that for every $i = 1, \dots, k{:}K_i = (s_i, t_i, d_i)$. By definition, the *fractional (general) routing problem* is the problem of finding k fractional flows denoted by sequence $\bar{F} = (F_1, F_2, \dots, F_k)$ such that F_i is a fractional flow of value d_i from s_i to t_i in graph G. Sequence $\bar{F}$ is called a *fractional solution* of the problem.

As recently defined, by applying the following constraint on the fractional solution, the problem will turn into the general routing problem with integral solution.

$$|F_i| = 1 \quad \forall i = 1, 2, \dots, k$$

Consequently, every fractional flow will only have one member in the form (p_i, d_i); i.e.:

$$F_i = \{(p_i, d_i)\} \quad \forall i = 1, 2, \dots, k \tag{1.16}$$

Similar to the general routing problem, we now define the flow function of the fractional solution.

Definition 1.16   Let $\bar{F}$ denote a solution of the fractional routing problem and let $G = (V, E)$ specify the problem graph. Function $f_{\bar{F}}^{(i)}: E \mapsto \mathbb{R}_{\geq 0}$ is the *flow function of the ith commodity* in solution $\bar{F}$ and is defined by the following equation:

$$f_{\bar{F}}^{(i)}(e) = \sum_{(p,x)\in F_i} \sum_{e\in p} x \quad \forall e \in E, \forall i = 1, 2, \dots, k \tag{1.17}$$

such that $\bar{F} = (F_1, F_2, \dots, F_k)$. The value of $f_{\bar{F}}^{(i)}(e)$ is called the *traffic flow* of the *i*th commodity through edge e in fractional solution $\bar{F}$.

Regarding equation (1.17), the flow function of the fractional solution can get the values between zero and d_i. In other words, despite the integral solution, function $f_{\bar{F}}^{(i)}(e)$ may output more than two values.

Now, consider the constrained version of the fractional routing problem, which is equivalent to the one with the integral solution. In this special case, considering the form of each fractional flow shown in equation (1.16), we can simplify the flow function in the following form:

$$f_{\bar{F}}^{(i)}(e) = \sum_{(p,x)\in\{(p_i,d_i)\}} \sum_{e\in p} x$$

$$= \sum_{e\in p_i} d_i = \begin{cases} d_i & e \in p_i \\ 0 & \text{otherwise} \end{cases} \quad \forall e \in E, \forall i = 1, 2, \dots, k$$

where d_i is the total flow value of F_i. As shown in this equation, $f_{\bar{F}}^{(i)}(e)$ outputs the same value as flow function $f_{\bar{p}}^{(i)}(e)$ does in the (integral) general routing problem.

Routing Cost of the Fractional Solution   For a fractional routing problem, function cost_e of the edge routing cost, function nrc_π of the network routing cost, and triple $\mathcal{E} = (\text{cost}_e, \text{nrc}_\pi, \bar{K})$ of the routing cost environment are defined exactly the same as the general routing problem. Additionally, our routing cost optimization problem can be extended to the problem of finding the optimized fractional solution, which incurs the minimum network routing cost among all the possible fractional solutions of the problem; i.e.:

$$\text{minimize}_{\bar{F}\in\mathcal{F}_{\bar{K}}}\{\text{nrc}_\pi(C_{\bar{F}}(\pi_1), C_{\bar{F}}(\pi_2), \dots, C_{\bar{F}}(\pi_{|E|}))\}$$

such that $\mathcal{F}_{\bar{K}}$ denotes the set of all the solutions of the fractional routing problem with commodities $\bar{K}$ and also the routing cost incurred in every edge e of the network is obtained by the following equation:

$$C_{\bar{F}}(e) = \text{cost}_e\left(f_{\bar{F}}^{(1)}(e), f_{\bar{F}}^{(2)}(e), \dots, f_{\bar{F}}^{(k)}(e)\right)$$

Assuming $\bar{F}_{\mathcal{E}}^{*}$ to be the optimized fractional solution, the minimum cost of routing the commodities is obtained by the following equation:

$$C_{\mathcal{E}}^{*} = \text{nrc}_\pi\left(C_{\bar{F}_{\mathcal{E}}^{*}}(\pi_1), C_{\bar{F}_{\mathcal{E}}^{*}}(\pi_2) \dots, C_{\bar{F}_{\mathcal{E}}^{*}}(\pi_{|E|})\right)$$

Obliviousness of the Routing Cost Environment   At this point, we have prepared the tools needed to define the fractional routing problem in an oblivious environment. This problem is the set of fractional routing problems that each one has different routing cost environments (similar to what was mentioned in Definition 1.11). The following definition presents the general form of the solution function of such routing problems that are defined in an oblivious routing cost environment.

Definition 1.17   Consider a fractional routing problem of graph $G = (V, E)$ and set $\mathbb{E}$ of the possible routing cost environments. For this problem, a *fractional solution* function is defined in the following form:

$$S: \mathbb{E} \mapsto \bigcup_{\mathcal{E} \in \mathbb{E}} \mathcal{F}_{\bar{K}_{\mathcal{E}}}$$

such that for every environment $\mathcal{E}$ that belongs to $\mathbb{E}$, $\bar{K}_{\mathcal{E}} = \Pi_3(\mathcal{E})$, set $\mathcal{F}_{\bar{K}_{\mathcal{E}}}$ contains all the fractional solutions of the routing problem with commodities $\bar{K}_{\mathcal{E}}$ in graph G; and for every environment $\mathcal{E} \in \mathbb{E}$, the following relation holds: $S(\mathcal{E}) \in F_{\bar{K}_{\mathcal{E}}}$.

In the above definition, solution $S(\mathcal{E})$ is called the solution suggested by function S for the fractional routing problem of graph G in the routing cost environment $\mathcal{E} \in \mathbb{E}$.

Moreover, in order to find the optimized solution function using the dynamic approach, we have to compute the sequence $\bar{F}_{\mathcal{E}}^*$ which minimizes the network routing cost in every possible routing cost environment $\mathcal{E} \in \mathbb{E}$; i.e.:

$$S_{\text{dynamic}}(\mathcal{E}) = \bar{F}_{\mathcal{E}}^* \quad \forall \mathcal{E} \in \mathbb{E}$$

Having established this, in the versatile approach of obtaining the solution function, we have to first define a *fractional versatile routing scheme* (like the integral version of the oblivious routing problem).

Definition 1.18 Consider a fractional routing problem of graph $G = (V, E)$ and set $\mathbb{E}$ of the possible routing cost environments. The *fractional versatile routing scheme* is defined as the following function:

$$\mathbb{S}: V^2 \mapsto \text{the set of all the possible fractional flows in } G$$

such that for every $u, v \in V$, $\mathbb{S}(u, v)$ determines a fractional flow of unit value, source vertex u, and target vertex v in graph G. The solution function associated with $\mathbb{S}$ is represented by $S_{\mathbb{S}}$. Assuming that $(\text{cost}_e, \text{nrc}_\pi, \bar{K})$ is an arbitrary member of $\mathbb{E}$ and sequence $\bar{K}$ is equal to $(K_1, K_2, \ldots, K_k)$, the following equation holds:

$$S_{\mathbb{S}}(\text{cost}_e, \text{nrc}_\pi, \bar{K}) = (F_1, F_2, \ldots, F_k)$$

such that:

$$F_i = \{(p, x \times \Pi_3(K_i)): (p, x) \in \mathbb{S}(\Pi_1(K_i), \Pi_2(K_i))\} \quad \forall i = 1, 2, \ldots, k$$

Moreover, the competitiveness ratio of fractional scheme $\mathbb{S}$ is defined like the ratio of the integral one:

$$\text{CR}(\mathcal{E}, \mathbb{S}) = \max_{\mathcal{E} \in \mathbb{E}} \frac{C_{\mathcal{E}}(S_{\mathbb{S}}(\mathcal{E}))}{C_{\mathcal{E}}(\bar{F}_{\mathcal{E}}^*)}$$

1.5 Summary and Outlook

In this chapter, a general overview of different network design optimization problems, and preliminary definitions were presented. In the first two sections, the basic and general version of the network routing problem was defined. Moreover, a general routing cost model was constructed and illustrated by two examples. This model can be used in many real-world examples of the network routing problem.

In the third section, the concept of obliviousness in network design was precisely defined. Oblivious flow in the general routing problem is an important issue in various network applications. The only practical way to solve many real-world examples of the general routing problem with oblivious flow is to use the versatile approach, which uses a versatile routing scheme.

Finally, in the fourth section, the concept of the fractional routing was compared with integral routing. We also showed that these two types of routing problems can have the same optimal solution in some circumstances.

Exercises

1. Consider p as a simple path in graph $G = (V, E)$. Show that for every triple of edges $e_1, e_2, e_3 \in p$, the following proposition holds:

$$((e_1 \neq e_2) \wedge (e_2 \neq e_3) \wedge (e_3 \neq e_1)) \leftrightarrow (e_1 \cap e_2 \cap e_3 = \varnothing)$$

This property of the simple path guarantees that there exists no repetitive vertex on the path.

2. Consider p as an arbitrary walk of no repetitive edge in graph $G = (V, E)$. Show that there is a set of edges like q such that $q \subseteq p$ and q specifies a simple path in G.

3. Consider the transportation example in Section 1.1.
   a. Find the solution function associated with the solution shown in figure 1.3.
   b. Prove that the edge routing cost function presented in equation (1.2) is concave.
   c. In the case that $a \ll 1$, is this function a separable cost function? If yes, what is the rrc(x) and weight of each edge?

4. Show how the single-source network routing problem is a special case of the general version.

5. Consider the general network routing problem. Prove that if we are concerned about the flow congestion in the vertices of the problem graph, the edge routing cost function will be the summation function and the nrc$_\pi$ will be in the following form:

$$\mathrm{nrc}_\pi(c(e_1), c(e_2), \ldots, c(e_{|E|})) = \frac{1}{2}\mathrm{max}_{v \in V}\sum_{\substack{e \in E \\ v \in e}} c(e)$$

6. Consider connected graph $G = (V, E)$ as a representation of a data network such that $|V| = n$. Moreover, assume a versatile routing problem of graph G that any two vertices can send data to each other through the network. In the case that the edge cost function of each edge is obtained by the following equation and considering that we are also interested in minimizing the total cost, find the set of all the possible cost environments of the problem. How many members does this set have?

$$\text{cost}_e(x) = w_e \cdot \text{rrc}(x)$$

Suggested Reading

Alon, N., B. Awerbuch, Y. Azar, N. Buchbinder, and J. S. Naor. 2004. A general approach to online network optimization problems. *Proceedings of the 15th Symposium on Discrete Algorithms, SODA '04*, 942–951.

Andrews, M. 2004. Hardness of buy-at-bulk network design. *Proceedings of the 45th FOCS*, 115–124.

Azar, Y., E. Cohen, A. Fiat, H. Kaplan, and H. Racke. 2003. Optimal oblivious routing in polynomial time. *Proceedings of the 35th STOC*, 383–388.

Gupta, A., M. T. Hajiaghayi, and H. Racke. 2006. Oblivious network design. *Proceedings of the Seventeenth Annual ACM-SIAM Symposium on Discrete Algorithms, SODA '06* 970–979.

Gupta, A., A. Kumar, and T. Roughgarden. 2003. Simpler and better approximation algorithms for network design. *Proceedings of the 35th STOC*, 365–372.

Srinivasagopalan, S. 2011. Oblivious buy-at-bulk network design algorithms. Doctoral dissertation, Louisiana State University, Baton Rouge, LA.

Srinivasagopalan, S., C. Busch, and S. S. Iyengar. 2011a. An oblivious spanning tree for single-sink buy-at-bulk in low doubling-dimension graphs, Louisiana State University, Baton Rouge, LA.

Srinivasagopalan, S., C. Busch, and S. S. Iyengar. 2011b. *Oblivious buy-at-bulk in planar graphs*, Workshop on Algorithmic Computing (WALCOM), IIT-Delhi, 18–20 February.

I Mathematical Foundation

Chapter 2   Hierarchical Routing Tools and Data Structures

Data structures are tools for storing and organizing data so that it can be used efficiently. There are a variety of data structures that are designed for different goals. The branch of data structures that use the *Divide and Conquer* technique to efficiently organize a large-scale collection of data in a scalable way are known as *hierarchical data structures*.

In the recent decades, during the world-wide process of globalization, the size of many real-world networks has been substantially increased. Subsequently, the graph representation of these networks has gradually become larger and more complicated. To deal with the routing problems defined in such large-size graphs, we need to have some well-designed data structures like hierarchical ones so that they prevent a huge growth in the cost of the routing algorithms.

In chapter 1, different versions of the routing problem in a network were presented thoroughly. We used the graphs as a common type of linked data structures to represent an arbitrary network. Graph representation of a network made it easy for us to formulate the routing problems; however, we need more complex linked data structures to use in the solving algorithms of the formulated problems. These data structures are preferable to be highly flexible to the graph size as it may vary from tens to millions nodes. Moreover, in order to solve the problems in oblivious routing cost environments, it is important to design data structures that are versatile to the changing environment. Finally, we need efficient tools to reduce the cost and time of routing computations.

In this chapter, our goal is to find the data structures that have these criteria. More specifically, we will focus on the hierarchical data organizing to make our design scalable. Here is a brief description of the hierarchical approach used in this chapter: we define different levels of abstraction for the problem graph. The most abstract model of the graph represents the first (highest) level of abstraction; while the lower levels, model the graph in a more detailed way; and the lowest level is the graph itself.

In the first section, we will present some preliminary definitions. In sections 2.2 and 2.5, we will examine two different data structures: (1) the hierarchical decomposition tree, which organizes data in a *top-down* manner, and (2) the hierarchical routing tree, which does it in a *bottom-up* fashion. In the next chapter, we will see how these tools can be applied to make efficient routing algorithms.

2.1 Preliminary Definitions

In this section, we define some preliminary terms that will be used in the rest of this chapter.

Here, we introduce the *weighted* undirected graph as an extension of undirected graphs specified in Definition 1.1.

Definition 2.1 Let $G' = (V, E)$ denote an undirected graph. For some function $w : E \mapsto \mathbb{R}^+$, triple $G = (V, E, w)$ is defined as the *weighted undirected graph* corresponding to G' where V, E, and w are called the vertex set, edge set, and the *weight function* of G, respectively. Moreover, the (simple) path between two vertices in G is defined as the path between them in G'. Also, G is connected if and only if G' is connected.

Here, we assume that any undirected graph (V, E) is equivalent to the weighted undirected graph (V, E, unit), where $\forall e \in E : \text{unit}(e) = 1$. In other words, every definition or claim about the weighted undirected graph can be extended to the unweighted graph using this equivalence. Note that in all of the following definitions, weighted undirected graphs and (unweighted) undirected graphs are simply referred to as weighted graphs and (unweighted) graphs, respectively.

Considering $G = (V, E, w)$ as a weighted graph, the length of simple path p (in graph G) is specified by function $\text{len}_G : 2^E \mapsto \mathbb{R}^+$, and is defined by the following equation:

$$\text{len}_G(p) = \sum_{e \in p} w(e)$$

Distance function $d_G : V^2 \mapsto \mathbb{R}^+$ of weighted graph $G = (V, E, w)$ is defined in the following form:

$$d_G(u, v) = \min_{p \in P(u,v)} \text{len}_G(p) \tag{2.1}$$

where $P(u, v)$ denotes the set of all the simple paths existing between vertices u and v in G. Note that if weighted graph G is replaced with its unweighted equivalent, i.e., if we assume that the weight function of graph G is the unit function, the length of a path is obtained by the following equation:

$$\text{len}_G(p) = \sum_{e \in p} \text{unit}(e) = \sum_{e \in p} 1 = |p|$$

Hence, the distance between any two vertices of a connected graph is *the length of the shortest path existing between them*. Additionally, if graph G is not connected, as set $P(u, v)$ will be empty for some $u, v \in V$, we need to redefine the distance function:

$$d_{G(u,v)} = \begin{cases} \min_{p \in P(u,v)} \text{len}_G(p) & P(u, v) \neq \varnothing \\ +\infty & \text{otherwise} \end{cases}$$

The *maximum value* of all the finite distances in a connected graph is called the *diameter of graph* G and is represented by diam(G):

$$\text{diam}(G) = \begin{cases} \max_{u,v \in V} d_G(u, v) & G \text{ is connected} \\ \text{undefined} & \text{otherwise} \end{cases}$$

Moreover, *an r-neighborhood of vertex* v or *ball* $B_G(v, r)$ in the weighted graph $G = (V, E, w)$ is defined by the following set:

$$B_G(v, r) = \{u \in V : d_G(u, v) \leq r\}$$

For any set $V' \subseteq V$, the *subgraph* of graph $G = (V, E, w)$ *induced by* V' is defined as graph $G_{V'} = (V', E', w)$ where:

$$E' = \{\{u, v\} : u, v \in V' \wedge \{u, v\} \in E\}$$

We now define the Δ-partition of an arbitrary weighted graph.

Definition 2.2 For any $\Delta \geq 0$, the Δ-partition of connected graph $G = (V, E, w)$ is a partition[1] of vertex set V into a number of subsets (which are called as the clusters) such that:

$$\forall C \in \Delta\text{-partition}(G): \max_{u,v \in C} d_G(u, v) \leq \Delta$$

By setting the Δ parameter of Definition 2.2 to different values, we will get various partitions of the graph vertex set. Four properties of the Δ-partition of weighted graph $G = (V, E, w)$ are as follows:

1. For a specific value of Δ, the Δ-partition of graph G is not unique.
2. For every $\Delta \geq \text{diam}(G)$, partition $\{V\}$ specifies a Δ-partition of G.
3. Assuming that $\Delta_1 \leq \Delta_2$, every Δ_1-partition of graph G is also a Δ_2-partition of G.
4. Assuming that C is a cluster of set Δ-partition(G), subgraph G_C is *not* necessarily connected.

In order to illustrate Definition 2.2, consider figure 2.1, which shows a weighted graph of diameter four represented by $\mathcal{G} = (V', E')$. In graph $\mathcal{G}$,

1. Partition of set A is a set of A subsets $\{A_i : i = 1, 2, \ldots, k\}$ such that $\bigcup_{i=1}^{k} A_i = A$, $A_i \neq \varnothing$ $\forall i = 1, 2, \ldots, k$, and $A_i \cap A_j = \varnothing$ $\forall i \neq j$.

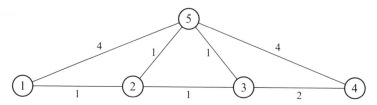

Figure 2.1 A connected weighted graph of diameter four.

set $X = \{\{1\}, \{2\}, \{3, 4, 5\}\} \subseteq 2^{V'}$ specifies a partition of the vertex set V'. Additionally, the maximum graph distance between any two vertices in set $\{3, 4, 5\} \in X$ is not greater than three ($d_G(3, 4) = 2 \leq 3$, $d_G(4, 5) = 3 \leq 3$, and $d_G(3, 5) = 1 \leq 3$). A similar proposition also holds for the other two members of set X as they only contain one vertex ($\{1\}$ and $\{2\}$).

Henceforth, set X is a 3-partition of graph $\mathcal{G}$. In a similar way, it can be shown that set $X' = \{\{1, 2, 3, 5\}, \{4\}\}$ is a 2-partition of graph $\mathcal{G}$:

$$\forall u, v \in \{1, 2, 3, 5\} : d_G(u, v) \leq 2$$

Regarding the listed properties of Δ-partitions, note that set X' can also be considered as a 3-partition of graph $\mathcal{G}$; however, set X is not a 2-partition of $\mathcal{G}$ as $d_G(4, 5) > 2$.

As another example, consider figure 2.2, which represents two different 4-partitions of a weighted graph. In each partition, the way that clusters are scattered over the graph and the subgraphs induced by them have been highlighted.

2.2   Hierarchical Decomposition Tree

We now introduce the *hierarchical decomposition tree* as an example of hierarchical routing data structures that use a top-down approach.

2.2.1   Hierarchical Decomposition Sequence

Let $G = (V, E, w)$ denote a weighted connected graph. The *hierarchical decomposition sequence* (HDS) $\bar{H}$ of graph G is defined as the sequence $\bar{H} = (H_0, H_1, \ldots, H_h)$ where:

- $h = \lceil \log_2 \text{diam}(G) \rceil$.
- $H_h = \{V\}$.
- For every $i = 0, 1, \ldots, h - 1$, H_i is a 2^i-partition.
- For every $i = 0, 1, \ldots, h - 1$, if C belongs to H_i, the following proposition holds:

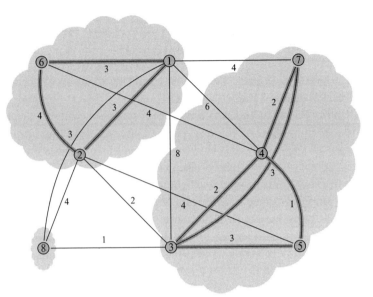

(a) Vertex set partition $\{\{8\},\{1,2,6\},\{3,4,5,7\}\}$

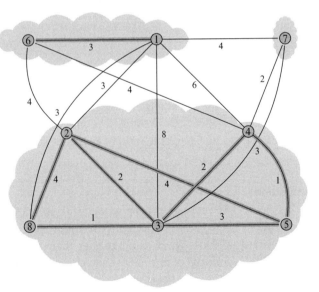

(b) Vertex set partition $\{\{7\},\{1,6\},\{2,3,4,5,8\}\}$

Figure 2.2 Two different 4-partitions of a weighted graph.

$$\exists C' \in H_{i+1} : C \subseteq C'$$

Partition H_i is called the ith *level partition* of $\bar{H}$.

Note that the HDS is only defined for connected graphs. In addition, there are $\lceil \log_2 \text{diam}(G) \rceil + 1$ partitions of set V in an HDS that are not necessarily distinct (as HDS is a sequence, not a set).

Each partition of a hierarchical decomposition sequence defines a level of abstraction in the graph. At the h level (the highest level), the whole graph's vertex set is covered by only one cluster. At level $h - 1$, the distance high threshold between the vertices of each cluster is equal to 2^{h-1} (which is half of the threshold in the upper level). Subsequently, in the lower level partitions, the aforementioned high threshold will decrease exponentially. This implies that there may exist more clusters in the lower levels of an HDS. In addition, as each cluster in the $i + 1$ level partition is the union of some clusters in the ith level partition (for every $i < h$), the number of clusters does not decrease in the lower levels. As a result, partitioning of the vertex set gradually becomes *more refined* as we make our way to the lower levels.

In figure 2.3, one of the possible hierarchical decomposition sequences of the given graph is shown. In general, the union of all the partitions of an HDS makes a *laminar family*[2] of the vertex set. An obvious HDS of any connected graph is found in the following form:

- $H_h = V$.
- $H_i = \{\{v\} : v \in V\} \quad \forall i = 0, 1, \ldots, h - 1$.

In the rest of this section, by graph we mean weighted connected graph. Any statement concerning the weighted graphs in this section is also extendable to the unweighted graphs (as (V, E) is equivalent to $(V, E, unit)$).

2.2.2   Tree Definition

As mentioned before, the HDS of a graph is a sequence of vertex set partitions. The ith partition (H_i) of the sequence contains a number of clusters that are more refined than those in H_{i-1} and coarser than the members of H_{i+1} (note from the definition of the HDS that clusters of the lower partitions are subsets of those in the higher partitions). This relation between the clusters of subsequent partitions leads us to define a laminar

---

2. Set $\mathcal{L} \subseteq 2^X$ is called a *laminar family* of set X if for every $A, B \in \mathcal{L}$, one of the following conditions holds: $A \subseteq B$, $B \subseteq A$, or $A \cap B = \emptyset$. Every partition of set X is a laminar family of X.

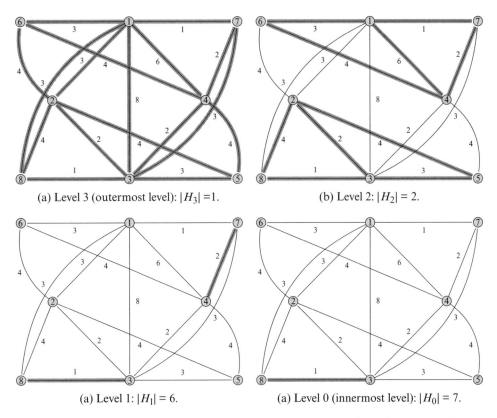

(a) Level 3 (outermost level): $|H_3| = 1$.

(b) Level 2: $|H_2| = 2$.

(a) Level 1: $|H_1| = 6$.

(a) Level 0 (innermost level): $|H_0| = 7$.

Figure 2.3 The hierarchical decomposition levels of a connected weighted graph of diameter eight. Note that the number of clusters in the lower level partitions is greater than or equal to the higher ones, i.e., $|H_3| \le |H_2| \le |H_1| \le |H_0|$.

tree[3] for every HDS of a graph. The formal definition of this tree is given as follows.

Definition 2.3 Let $\bar{H} = (H_0, H_1, \ldots, H_h)$ denote an HDS over the graph $G = (V, E, w)$. Tree $T_G(\bar{H}) = (V_T, E_T)$ of root H_h is called the *Hierarchical Decomposition Tree* (HDT) of HDS $\bar{H}$ if the following equations hold:

$$V_T = \{(C, i) : C \in H_i, i = 0, 1, \ldots, h\}$$

$$E_T = \{\{(C_1, i), (C_2, i+1)\} : C_1 \subseteq C_2, i = 0, 1, \ldots, h-1\}$$

3. Tree T of root r is a connected acyclic graph (V, E) with a distinguished vertex $r \in V$ called the *root*. Considering $\mathcal{L}$ as a laminar family of X, *laminar tree* $T = (V, E)$ of root r and family $\mathcal{L}$ is a tree such that every vertex $v \in V$ corresponds to set $S_v \in \mathcal{L}$, and for every edge $\{u, v\} \in E$, if $d_T(u, r) > d_T(v, r)$, $S_v \subseteq S_u$.

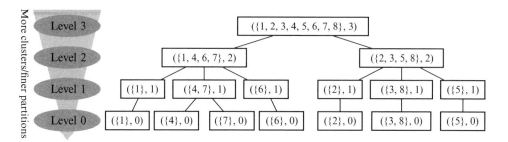

Figure 2.4 Hierarchical decomposition tree of the HDS shown in figure 2.3.

In the above relations, $h = \lceil \log_2 \text{diam}(G) \rceil$. Additionally, for any tree vertex $(C, 0) \in V_T$, C is called a *basic cluster*.

Concerning the above definition, in the hierarchical decomposition tree of an HDS, every vertex is an ordered pair of a cluster and a number that specifies the level of the HDS partition containing the cluster. Additionally, each edge of the HDT connects a pair of vertices if their corresponding clusters belong to the partitions of the successive levels and also one cluster contains the other. Note that in the case that graph G is unweighted, the induced subgraph of every basic cluster in an HDT is a complete graph.[4]

For an instance of the HDT, see figure 2.4, which shows the hierarchical decomposition tree of the HDS presented in figure 2.3. Note that in the lower levels, the number of clusters increases and the partitions become finer.

2.3   Hierarchical Independence Tree Type-1

In this section, we introduce the *Hierarchical Independence Tree* (HIT) as a hierarchical tool to organize the routing related data in the bottom-up fashion. Before formally defining the HIT, the independence relation between the vertices of a graph will be addressed thoroughly. Furthermore, we will discuss how the HIT partitions a graph by comparing it with the HDT partitioning.

2.3.1   Independent Set of Vertices

Let $G = (V, E, w)$ denote a connected graph. Set $I \subseteq V$ as an *r-independent set* of graph G if:

---

4. Graph $G = (V, E)$ is complete if for every $u, v \in V$, $d(u, v) = 1$. Complete graphs are represented by symbol K_i where i is the cardinality of its vertex set.

$$\forall u, v \in I : d_G(u, v) \geq r$$

Now, consider set $J \subseteq V$ as an r-independent set of G. Set J as the *maximum r-independent set* of G if for every r-independent set J' of graph G, $|J| \geq |J'|$. Similarly, the r-independent set $K \subseteq V$ is a *maximal r-independent set* of G if the following condition holds:

$$\forall K' \subseteq V : (K' \text{ is an } r\text{-independent set} \land K \subseteq K') \rightarrow (K = K')$$

In addition, assuming that $I \subseteq V$ is an r-independent set of graph G, set J is a *maximal r-independent set of graph G including set I* if J specifies a maximal r-independent set and $I \subseteq J$.

Concerning the above definitions, the maximum r-independent set of graph G has the maximum cardinality among all of the r-independent sets of G. For every $r_1, r_2 \in \mathbb{R}_{\geq 1}$, assuming that $r_2 \geq r_1$, the maximum r_2-independent set is not larger than the maximum r_1-independent set, because every r_2-independent set is also an r_1-independent set (in the extreme case, the maximum one-independent set of G is its vertex set, which has the maximum possible number of vertices). In a similar manner, for every maximal r_1-independent set of G, there is no other r_2-independent set that contains it thoroughly.

To illustrate the independent sets, we use the family of two-dimensional grid graph.[5] As shown in figure 2.5, the maximum 4-independent set and the maximal 4-independent set, including set $\{1, 2, 3, 4\}$ of vertices, have been specified for graph $G_{9\times9}$.

Assume $G = (V, E, w)$ is a connected graph and value $h = \lceil \log_2 \text{diam}(G) \rceil$. Considering vertex $s \in V$ as a distinguished vertex, sequence $\bar{I}_s = (I_0, I_1, I_2 \ldots, I_h)$ is called a *hierarchical independent sequence* (HIS) of source vertex s and graph G if both of the following conditions are met:

- $I_h \{s\}$.
- $\forall i = 0, 1, \ldots, h - 1$, I_i is a maximal 2^i-independent set of G including I_{i+1}.

Some basic properties of the hierarchical independence sequence are as follows:

1. Assuming that $w(u, v) \geq 1$ for every $\{u, v\} \in E$, I_0 contains all the graph vertices, i.e., $I_0 = V$.
2. $I_h = \{s\}$ is *not* necessarily a maximal 2^h-independent set.
3. The HIS of a graph is not unique.

5. Two-dimensions grid graph $G_{n\times m}$ is a connected unweighted undirected graph (V, E) such that $V = \{v_{ij} : i = 1, 2, \ldots, n, j = 1, 2, \ldots, m\}$ and $E = \{\{v_{ij}, v_{i(j+1)}\} : i = 1, 2, \ldots, n, j = 1, 2, \ldots, m - 1\} \cup \{\{v_{ij}, v_{i(j+1)}\} : i = 1, 2, \ldots, n - 1, j = 1, 2, \ldots, m\}$.

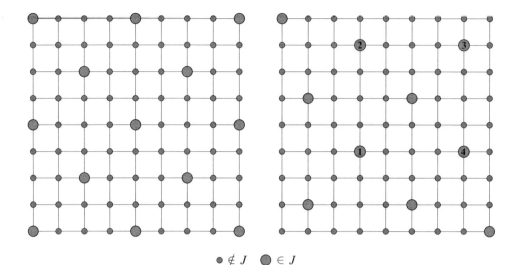

● ∉ *J*   ⬤ ∈ *J*

Figure 2.5 Example of grid graph $G_{9\times9}$. The left graph specifies set J as a maximal 4-independent set of the graph (in this case $|J| = 13$). The right one specifies J as a maximal 4-independent set of $G_{9\times9}$ including set $I = \{1,2,3,4\}$ (in this case, $|J| = 10$). Note that set I is a 4-independent set of the graph.

4. For every $i, j = 0, 1, \ldots, h$, the following proposition is true:

$$(i < j) \leftrightarrow (|I_i| \geq |I_j|)$$

5. For every $i = 0, 1, \ldots, h$, s belongs to set I_i.

Figure 2.6 schematically depicts an HIS of the shown source vertex in grid graph $G_{5\times5}$.

2.3.2   HIT Definition and Properties

Now, we are ready to define the hierarchical independence tree type-1:

Definition 2.4   Let $G = (V, E, w)$ denote a connected graph and $h = \lceil \log_2 \text{diam}(G) \rceil$. In addition, assume that $\bar{I}_s = (I_0, I_1, I_2, \ldots, I_h)$ is a hierarchical independent sequence of source vertex $s \in V$ in graph G. Tree $T^s = (V_{T^s}, E_{T^s})$ of root (s, h) is the *hierarchical independence tree (HIT) type-1* associated with HIS $\bar{I}_s$ if:

$$V_{T^s} = \{(v, i) : v \in I_i, i = 0, 1, \ldots, h\} \tag{2.2}$$

and $E_{T^s} = E_1 \cup E_2$ where:

$$E_1 = \{\{(s, h), (v, h-1)\} : v \in I_{h-1}\} \tag{2.3}$$

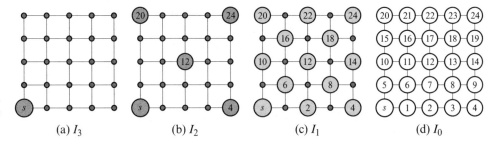

(a) I_3 (b) I_2 (c) I_1 (d) I_0

Figure 2.6 Considering $(I_0, I_1, ..., I_3)$ as a hierarchical independent sequence of source vertex s in grid graph $G_{5\times5}$, sets I_0, I_1, I_2 and I_3 have been specified in the above figures $(I_0 = V)$. In each figure, the vertices included in the corresponding maximal 2^i-independent set are gray. Note that $I_3 \subseteq I_2 \subseteq I_1 \subseteq I_0$.

and E_2 consists of the edges in the form $\{(u, i + 1), (v, i)\}$, such that for every $i = 0, 1, ... , h - 2$, vertex v belongs to set I_i, and $(u, i + 1)$ is an arbitrarily chosen member of set $\mathrm{Par}_1((v, i))$. Function $\mathrm{Par}_1 : V_{T^s} \mapsto 2^{V_{T^s}}$ is called the *parent function type-1* and is defined in the following form:

$\mathrm{Par}_1((v, i))$

$$= \left\{ (u, i+1) : u \in I_{i+1} \cap B_G\left(v, 2^{i+1} - 1\right), d_G(u, s) = \min_{\substack{x \in I_{i+1} \cap \\ B_G\left(v, 2^{i+1}-1\right)}} d_G(x, s) \right\} \quad (2.4)$$

From Definition 2.4, each tree vertex is an ordered pair of some vertex in I_i and index $i = 0, 1, ... , h$. Vertex set $L_i = \{(v, i) : v \in I_i\}$ is called the set of the ith level vertices[6] of the HIT, for every $i \leq h$ (where h specifies the tree height). Moreover, the edge set of the HIT is divided into two partitions E_1 and E_2. E_1 consists of edges that connect the $h - 1$ level vertices to the source vertex (s) which is the only vertex in level h, and E_2 members connect pairs of vertices together in a way that for every $i = 0, 1, ... , h - 2$ and $v \in I_i$, the parent[7] of vertex (v, i) will be $(u, i + 1)$ such that u is the closest vertex to s in G that holds the following two conditions:

- $u \in I_{i+1}$.
- $d_G (u, v) \leq 2^{i+1} - 1$ or $u \in B_G (v, 2^{i+1} - 1)$ (the distance high threshold between child and parent).

6. In tree $T = (V, E)$ of root $r \in V$, a number is mapped to each vertex $v \in V$ as its level, represented by $\mathrm{level}_T (v)$. The level of root r is defined as $\max_{v \in V} d_T (r, v)$. The level of every other vertex $v \neq r$ is then equal to $\mathrm{level}_T (r) - d_T (r, v)$. The tree height is defined as $\mathrm{level}_T (r)$.

7. In tree $T = (V, E)$, for every edge $\{u, v\} \in E$, if $\mathrm{level}_T (u) > \mathrm{level}_T (v)$, v is called a child of u, and u is a parent of v.

To satisfy these conditions, the parent of every vertex at level $i \leq h - 2$ has to be chosen among the members of the output set of function Par_1. Lemma 2.2 will show that Par_1 output is not empty for every level $i \leq h - 2$ vertex; however, its output may contain more than one vertex, which leads to different HITs corresponding to one hierarchical independent sequence.

Figure 2.7 represents an HIT corresponding to the hierarchical independent sequence of graph $G_{5\times5}$ shown in figure 2.6. In this example, as an illustration of function Par_1, we compute $Par_1((22,1))$. Among all the five members of I_2, vertices 12, 20, and 24 are close enough to vertex 22 (their distance from vertex 22 is not greater than the threshold $2^2 - 1 = 3$). Additionally, vertices 12 and 20 are closer to s than vertex 24 (see figure 2.8); however, as both of them are equally far away from vertex s, set $Par_1((22,1))$ contains both of them:

$$Par_1((22, 1)) = \{(20, 2), (12, 2)\}$$

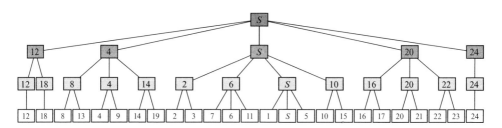

Figure 2.7 Brief representation of the HIT type-1 corresponding to the hierarchical independent sequence of graph $G_{5\times5}$ shown in figure 2.6. Each vertex of the HIT is an ordered pair represented by a rectangle of the figure in a way that its first element is denoted by the rectangle label and the second one is specified as the level of the rectangle in the shown tree.

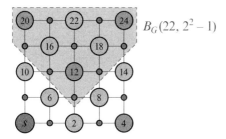
$B_G(22, 2^2 - 1)$

Figure 2.8 Depiction of the way of computing $Par_1((22, 1))$. Note that set $B_G(22, 2^2 - 1) \cap I_2$ is equal to $\{20, 12, 24\}$.

Note that in the depicted HIT, for every $i = 1, 2, \ldots, h$, vertex (v, i) has a child in the form $(v, i - 1)$. Generally, this property is true for an HIT type-1 of every HIS $\bar{I}_s = (I_0, I_1, \ldots, I_h)$ of the connected graphs.

Lemma 2.1 Assume HIS $\bar{I}_s = (I_0, I_1, \ldots, I_h)$ of source vertex $s \in V$ and connected graph $G = (V, E, w)$. Also, consider T^s as an HIT type-1 corresponding to $\bar{I}_s$. Every vertex (v, i) in tree T^s has a child in the form $(v, i - 1)$ where $i = 1, 2, \ldots, h$.

Proof. To prove this claim, we show that for an arbitrary tree vertex (v, i), vertex $(v, i - 1)$ exists and $\{(v, i), (v, i - 1)\}$ specifies a tree edge for every $i = 1, 2, \ldots, h$. As (v, i) is a vertex of the HIT, regarding equation (2.2), v must belong to I_i. Furthermore, since $I_i \subseteq I_{i-1}$, v also belongs to I_{i-1}. As a result, $(v, i - 1)$ is an HIT vertex (proof of existence).

Now, we show that vertex $(v, i - 1)$ is connected to (v, i), for every $i = 1, 2, \ldots, h$. In the case that $i = h$, as I_i is a subset of I_{i-1}, tree T^s has vertex $(s, h - 1)$ at the $h - 1$ level; on the other hand, regarding equation (2.3), all the vertices of tree T^s at the $h - 1$ level are connected to the root $((s, h))$. This implies that the proof is complete for this case.

If $i = 1, 2, \ldots, h - 1$, assume that vertex (u, i) is the parent of $(v, i - 1)$ in tree T^s; hence, we deduce that $(u, i) \in Par_1((v, i - 1))$. Concerning equation (2.4), vertex (u, i) must belong to set $B_G(v, 2^i - 1)$, or:

$$d_G(u, v) \le 2^i - 1 \tag{2.5}$$

In addition, since $u \in I_i$, $v \in I_i$, and I_i is a 2^i-independent set of G, $d_G(u, v) \ge 2^i$ unless $u = v$. Subsequently, regarding equation (2.5), u and v are the same and vertex (v, i) is the parent of $(v, i - 1)$ in HIT T^s. ∎

Before presenting a lemma that guarantees the existence of the hierarchical independence tree associated with an arbitrary HIS of a connected graph, we present some shortened notation here. Considering tree $T = (V, E)$ of root $s \in V$, the kth parent of node $v \in V$ is represented by $parent_k(v)$ and inductively defined as below:

- $\mathcal{P}_0(v)$ is v.
- $\mathcal{P}_{k+1}(v)$ is the parent of $\mathcal{P}_k(v)$ in tree T.

Lemma 2.2 Consider a connected unweighted graph $G = (V, E)$ and HIS $\bar{I}_s$ of source $s \in V$ in G. There exists an HIT type-1 corresponding to sequence $\bar{I}_s$.

Proof. Consider the definition of hierarchical independence tree $T^s = (V_{T^s}, E_{T^s})$. We first prove that for every $i = 0, 1, \ldots, h - 2$, if vertex (v, i) belongs to set V_{T^s}, $Par_1((v, i))$ will not be empty (assuming that $h = \lceil \log_2 diam_G \rceil \ge 2$). Then, we use this result to prove the existence of tree T^s.

Since $(v, i) \in V_{T^s}$, v is a vertex in I_i. On the other hand, as $I_{i+1} \subseteq I_i$, two possibilities exist: either $v \in I_{i+1}$ or $v \in I_i - I_{i+1}$. In the earlier case, concerning Lemma 2.1, vertex $(v, i+1)$ is the parent of (v, i). Consequently, according to the definition of the HIT type-1, $(v, i + 1) \in \mathrm{Par}_1((v, i))$ and $\mathrm{Par}_1((v, i)) \neq \emptyset$.

In the latter case, $v \in I_i$, but $v \notin I_{i+1}$. By contradiction, assume that $\mathrm{Par}_1((v, i)) \neq \emptyset$. Henceforth, regarding equation (2.4), $I_{i+1} \cap B_G(v, 2^{i+1} - 1) \neq \emptyset$. This implies that:

$$x \notin B_G\left(v, 2^{i+1} - 1\right) \rightarrow d_G(v, x) > 2^{i+1} - 1$$
$$\overset{(*)}{\rightarrow} d_G(v, x) \geq 2^{i+1} \quad \forall x \in I_{i+1} \tag{2.6}$$

Note that the last deduction (*) is made because of the assumption that graph G is unweighted (in an unweighted graph, the distance function always outputs an integer number).

Since I_{i+1} is a maximal 2^{i+1}-independent set, according to equation (2.6), set $I_{i+1} \cup \{v\}$ is also a 2^{i+1}-independent set. As $v \notin I_{i+1}$, we obtain $|I_{i+1} \cup \{v\}| > |I_{i+1}|$, which contradicts the assumption that I_{i+1} is a maximal 2^{i+1}-independent set.

Now, we need to prove that graph $T^s = \left(V_{T^s}, E_{T^s}\right)$ is an acyclic connected graph where V_{T^s} and E_{T^s} are defined in Definition 2.4. To do this, we should show that for every $u, v \in V_{T^s}$ ($u \neq v$), there is a path between u and v in T^s (connectivity) and this path is unique (acyclic).

Consider (u, i) and (v, j) as two arbitrary vertices of T^s where $(u, i) \neq (v, j)$. Without loss of generality, we assume that $i \geq j$. We continue the proof in these two possible cases: $i = j$ and $i > j$.

First, assume that $i = j$. In this case, regarding Definition 2.4, the $h - i$ parent of (u, i) and (v, j) is (s, h) (this is easy to show by induction); i.e.:

$$\mathcal{P}_{h-i}((u, i)) = \mathcal{P}_{h-i}((v, j))$$

Consider the smallest integer $k \leq h - i$ that makes this equation true: $\mathcal{P}_k((u, i)) = \mathcal{P}_k((v, j))$. Subsequently, for every $c = 0, 1, \ldots, k - 1$, $\mathcal{P}_c((u, i)) \neq \mathcal{P}_c((v, j))$. As a result, set $p = p_1 \cup p_2$ specifies a walk of no repetitive edge[8] from (u, i) to (v, j) in graph T^s where:

$$p_1 = \{\{\mathcal{P}_c((u, i)), \mathcal{P}_{c+1}((u, i))\} : c \leq k - 1\}$$

and

$$p_2 = \{\{\mathcal{P}_c((v, i)), \mathcal{P}_{c+1}((v, i))\} : c \leq k - 1\}$$

Note that p_1 is a walk from (u, i) to vertex $\mathcal{P}_k((u, i))$ in graph G and p_2 is a walk from (v, i) to vertex $\mathcal{P}_k((v, i))$, which is equal to $\mathcal{P}_k((u, i))$.

---

8. A walk of no repetitive edge was defined in section 1.1.

Consequently, if $i = j$, there is a walk (and also a simple path) that connects (u, i) and (v, j).

In the latter case, we only need to connect vertex (v, j) to its $i - j$ parent using the following walk, and then use the result of the first case to connect vertices $\mathcal{P}_{i-j}(v, j)$ and (u, i) together.

$$p' = \{\{\mathcal{P}_c((v, j)), \mathcal{P}_{c+1}((v, j))\} : c \le i - j - 1\}$$

Hence, in the case that $i > j$, walk $p \cup p'$ connects vertices (u, i) and (v, j) together. This implies that for every two vertices in graph T^s, there exists a path between them (proof of connectivity).

Again, consider (u, i) and (v, j) as two arbitrary vertices of T^s such that $(u, i) \ne (v, j)$ and $i \ge j$. Also, let A be an arbitrary path between these two vertices. Here, we prove by induction on j that for every $j = 1, 2, \dots, i$, if e denotes the edge connecting (v, j) to one of its children, $e \notin A$. In the case that $k = 1$, assume that $e \in A$. Regarding the path definition in chapter 1, $A = \{e\} \cup A'$, where A' is some path from a vertex in the form $(w, 0)$ to vertex (u, i) such that $e \notin A'$. As e is the only edge connecting $(w, 0)$ to other vertices of T^s, there is no path A' (contradiction). Assuming that children of (v, k) do not participate in path A for every $k = 1, 2, \dots, i - 1$ (induction hypothesis), we prove that the same statement is true for $k + 1$. Again, by contradiction, assume that $A = \{e\} \cup A'$ where e connects $(v, k + 1)$ to one of its children in the form (w, k), and A' denotes a path from (w, k) to (u, i) such that $e \notin A'$. In addition, according to induction hypothesis, none of the (w, k) children are crossed by A'; as a result, there is no edge in A' that connects (w, k) to any other vertex in T^s (contradiction).

From the above discussion, we conclude that for every $j = 0, 1, \dots, i$, the path from (v, j) to (u, i) contains the parent of (v, j), not its children (in the case that $j = 0$, vertex (v, j) has no children). Hence, regarding the path definition, we have:

$$A = \{\{\mathcal{P}_c((v, j)), \mathcal{P}_{c+1}((v, j))\} : c \le i - j\} \cup B$$

where B is a path from $(x, i+1) = \mathcal{P}_{i-j+1}((v, j))$ to (u, i). Assume that m is the smallest integer that holds this equation: $\mathcal{P}_m((x, i+1)) = \mathcal{P}_{m+1}((u, i))$ (we know with certainty that $m \le h - i$). Now, by induction on n, we prove that path B contains edge $e_n = \{\mathcal{P}_n((x, i+1)), \mathcal{P}_{n+1}((x, i+1))\}$ for every $n \le m - 1$. In the case that $n = 0$, if $e_n \notin B$, some edge between $(x, i + 1)$ and one of its children, say (y, i) belongs to B. As vertex $(x, i + 1)$ is not the parent of (u, i), it is the case that $(y, i) \ne (u, i)$, and according to the previous results, $\{(y, i), (x, i + 1)\} \notin B$, which means that vertex $(x, i + 1)$ is not crossed by B (contradiction). This implies that:

$$B = \{e_0\} \cup B_1$$

where B_1 is a path from $\mathcal{P}_1((x, i+1))$ to (u, i).

Here, we assume that $B = \{e_r : r = 0, 1, \ldots, k\} \cup B_{k+1}$, for every $k < m$, where B_k is a path from $\mathcal{P}_k((x, i+1))$ to (u, i). By contradiction, assume that e_{k+1} does not belong to path B_{k+1}. Subsequently, there exists an edge like e' which connects $\mathcal{P}_{k+1}((x, i+1))$ to one of its children, say $(z, i+k+1)$, such that $e' \notin B_{k+1}$; i.e.:

$$B_{k+1} = \{e'\} \cup B'$$

where B' is a path from $(z, i+k+1)$ to (u, i). As $(z, i+k+1)$ is not the parent of parents of (u, i), by similar inductive reasoning, we can show that B' does not contain any edge connecting $(z, i+k+1)$ to its children. Consequently, there would be no edge in B' connecting $(z, i+k+1)$ to any other vertex in T^s (contradiction). As a result,

$$B = \{e_r : r = 0, 1, \ldots, m-1\} \cup B_m$$

where B_m is a path from $\mathcal{P}_m((x, i+1)) = \mathcal{P}_{m+1}((u, i))$ to (u, i). With similar reasoning, we obtain the following equation:

$$B = \{e'_r : r0, 1, \ldots, m\} \cup B'_m$$

where $e'_r = \{\mathcal{P}_r((u, i)), \mathcal{P}_{r+1}((u, i))\}$ and B'_m is a path from $\mathcal{P}_{m+1}((u, i))$ to $(x, i+1)$. Consequently, the path from (v, j) to (u, i) is unique and can be written in the following form:

$$A = \{\{\mathcal{P}_c((v, j)), \mathcal{P}_{c+1}((v, j))\} : c \le i - j\} \cup \{e'_r : r \le m\} \cup \{e_r : r \le m-1\} \quad \blacksquare$$

2.3.3   Induced Partitions of HIT

In section 2.2, we addressed the hierarchical decomposition trees and the vertex set partitions of an arbitrary connected graph were created using the HDTs. Here, we also focus on another type of partitioning which is induced by the HIT type-1. We now present further shortened notation that will be used for the rest of the discussion.

Considering tree $T = (V, E)$ of root $s \in V$, the unique path between any two vertices of tree $u, v \in V$ is shown as $p_T(u, v)$. In addition, T_v represents the *subtree*[9] of tree T and is rooted at vertex $v \in V$. Additionally, the set of all the *leaves*[10] of tree T is denoted by $T \cdot leaves$.

---

9. $T' = (V', E')$ is called a subtree of tree $T = (V, E)$ rooted at $x \in V$ if graph T' is a subgraph of T induced by $V' = \{v \in V : \exists i, v = \mathcal{P}_i(x)\}$.

10. Assuming $T = (V, E)$ is a tree rooted at $s \in V$, vertex $v \in V$ is a tree leaf if and only if $\forall u \ne v : p_T(s, v) \nsubseteq p_T(s, u)$, where $p_T(u, v)$ denotes the unique path between every two vertices $u, v \in V$ in tree T.

Lemma 2.3 Consider HIT type-1 $T^s = (V, E)$ and the arbitrary vertex (r, i) $\in V$. Also, let $T' = (V', E')$ denote subtree $T^s_{(r,i)}$. Assuming that $(u, j) \in V'$, the following equation holds:

$$p_{T^s}((u, j), (r, i)) = p_{T'}((u, j), (r, i))$$
$$= \{\{\mathcal{P}_c((u, j)), \mathcal{P}_{c+1}((u, j))\} : c \leq i - j - 1\} \tag{2.7}$$

In addition, there is a walk of no repetitive edge in T^s and T' in the following form (let $(u', j') \in V'$):

$$p_{T^s}((u, j), (u', j')) = p_{T^s}((u, j), (r, i)) \, \Delta \, p_{T^s}((r, i), (u', j')) \tag{2.8}$$

Proof. This lemma is directly concluded from Lemma 2.2 and the subtree definition. ∎

Lemma 2.4 Let $T^s = (V_{T^s}, E_{T^s})$ denote a hierarchical independence tree type-1 in graph $G = (V, E)$. For every tree vertex $(r, i) \in V_{T^s}$, an arbitrary tree leaf of subtree $T^s_{(r,i)}$ is in the form $(l, 0)$ and:

$$d_G(l, r) < 2^{i+1} \tag{2.9}$$

Proof. Let (l, j) be a leaf of tree T'. At first, we prove that $j = 0$; then, we prove inequality (2.9).

By contradiction, assume that $j \neq 0$. Subsequently, according to Lemma 2.1, (l, j) has a child of the form $(l, j - 1)$ (in both T^s and T'). This implies that $p_{T'}((l, j-1), (r, i)) = p_{T'}((l, j), (r, i)) \cup \{\{(l, j-1), (l, j)\}\}$. As a result, $p_{T'}((l, j), (r, i)) \subseteq p_{T'}((l, j-1), (r, i))$ which contradicts the assumption that (l, j) is a leaf of T'. Hence, every leaf of T' is in the form $(l, 0)$ where l is some vertex in G.

Consider vertex $(l, 0) \in T'$·leaves. According to equation (2.7), since T' is a subtree of T^s rooted at (r, i),

$$p_{T^s}((l, 0), (r, i)) = \{\{\mathcal{P}_c((l, 0)), \mathcal{P}_{c+1}((l, 0))\} : c = 0, 1, \ldots, i-1\}$$

For every $c = 0, 1, \ldots, i - 1$, we map edge $\{\mathcal{P}_c((l, 0)), \mathcal{P}_{c+1}((l, 0))\}$ to path p_c which is defined in this way: $p_c \subseteq V$ is the shortest path in G from $\Pi_1(\mathcal{P}_c((l, 0)))$ to $\Pi_1(\mathcal{P}_{c+1}((l, 0)))$.

By this mapping, *the projection of path $p_{T^s}((l, 0), (r, i))$ on graph G* is obtained by the following formula.

$$p = \bigcup_{c=0}^{i-1} p_c \tag{2.10}$$

Concerning equation (2.10), $|p|$ is equal to $\sum_{c=0}^{i-1} |p_c|$. Since p_c is the shortest path from $\Pi_1(\mathcal{P}_c((l, 0)))$ to $\Pi_1(\mathcal{P}_{c+1}((l, 0)))$ in G, regarding the definition of Par_1 in equation (2.4), we conclude that:

$$|p_c| = d_G(\mathcal{P}_c((l, 0)), \mathcal{P}_{c+1}((l, 0)))$$
$$\leq 2^{c+1} - 1$$
$$\leq 2^{c+1}$$

Subsequently,

$$|p| = \sum\nolimits_{c=0}^{i-1} |p_c| < \sum\nolimits_{c=0}^{i-1} 2^{c+1} \leq 2^{i+1} - 2 < 2^{i+1}$$

Path p may cross itself, which results in the creation of cycles. By removing those edges that are participating in p cycles, we get simple path p' from $v_0 = l$ to $v_i = r$ in G with less cardinality than p. Finally, we obtain the following inequality:

$$d_G(l, r) \leq |p'| \leq |p| < 2^{i+1} \qquad \blacksquare$$

The hierarchical independence tree of height h in graph $G = (V, E)$ induces $h + 1$ partitions of set V. These partitions are obtained in the way specified by Definition 2.5.

Definition 2.5   Let $T^s = (V_{T^s}, E_{T^s})$ denote an HIT type-1 of HIS $I_s = I_0, I_1, \ldots, I_h$ in graph $G = (V, E)$. For every $i = 0, 1, \ldots, h$, the level-i induced partition of tree T^s is denoted by $\mathcal{V}_i$ and defined in the following way:

$$\mathcal{V}_i = \left\{ \mathcal{L}_{T^s}(v, i) : v \in I_i \right\}$$

where $\mathcal{L}_{T^s}(v, i) = \left\{ u : (u, 0) \in T^s_{(v,i)}.\text{leaves} \right\}$.

In Definition 2.5, note that for any two members u_1 and u_2 of $\mathcal{L}_{T^s}(v, i)$, the following inequality holds (triangle inequality):

$$d_G(u_1, u_2) \leq d_G(u_1, v) + d_G(v, u_2)$$

Moreover, regarding Lemma 2.4, we obtain:

$$d_G(u_1, v) + d_G(v, u_2) < 2^{i+1} + 2^{i+1}$$

Consequently, for every vertex $v \in I_i$, the following proposition is true:

$$\forall u_1, u_2 \in \mathcal{L}_{T^s}(v, i) : d_G(u_1, u_2) < 2^{i+2} \qquad (2.11)$$

Equation (2.11) shows that in induced partition $\mathcal{V}_i$, the vertices belonging to cluster $\mathcal{L}_{T^s}(v, i)$ become exponentially closer to each other as we go to the lower levels (with smaller i). This property implies that set $\mathcal{L}_{T^s}(v, i)$ is a 2^{i+2}-partition of the vertex set of graph G. In the end of chapter exercises, you will be asked to prove that the sequence $(\mathcal{V}_0, \mathcal{V}_0, \mathcal{V}_0, \mathcal{V}_1, \mathcal{V}_2, \ldots, \mathcal{V}_{h-3}, \{V\})$ specifies an HDS of graph G.

Similar to the HDS, in the induced partitions, cluster $\mathcal{L}_{T^s}(v, i)$ does not necessarily induce a connected subgraph in G. Figure 2.9 represents the level-1 and level-2 induced partitions of the HIT type-1 depicted in figure 2.7.

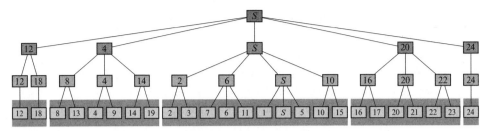

(a) The way that level-1 and level-2 partitions are induced by HIT type-1 of figure 2.7.

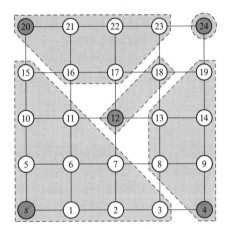

(b) Level-2 induced partition of HIT
shown in (a).

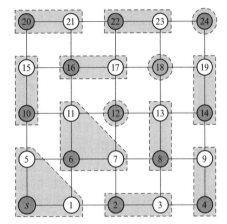

(c) Level-1 induced partition of HIT
shown in (a).

Figure 2.9 Induced level-1 and level-2 partitions of HIT type-1 shown in figure 2.7.

2.4 Hierarchical Independence Tree Type-2

The second type of the hierarchical independence trees is defined in a similar way as the first one was defined. The only difference is that the parent function is defined in a different way than what was mentioned in equation (2.4). Here is the definition of function $Par_2 : V_{T^s} \mapsto 2^{V_{T^s}}$:

$$
Par_2((v, i))
$$
$$
= \left\{ (u, i+1) : u \in I_{i+1} \cap B_G\left(v, 2^{i+2} - 2\right), d_G(u, s) = \min_{\substack{x \in I_{i+1} \cap \\ B_G(v, 2^{i+2} - 2)}} d_G(x, s) \right\} \quad (2.12)
$$

Note that the distance high threshold between child and parent at the ith level is $2^{i+2} - 2$, which is two times more than type-1's. Note that concerning equation (2.12), we obtain the following equation:

$$\text{Par}_2((v, h-1)) - \{(s, h)\} \quad \forall v \in I_{h-1}$$

This means that every vertex at the $h-1$ level is connected to the root as its child (like the HIT type-1).

Lemma 2.5 For every connected unweighted graph $G = (V, E)$ and HIS $\bar{I}_s$ of source $s \in V$ in G, there is an HIT type-2 corresponding to $\bar{I}_s$.

Proof. Just like the existence proof of the HIT type-1, we first prove that:

$$\forall v \in I_i : \text{Par}_2((v, i)) \neq \varnothing;$$

and then, we must show that T^s is an acyclic connected graph.

Let v be some vertex in set I_i. If $d_G(v, s) \leq 2^{i+2} - 2$, regarding equation (3.12), vertex $(s, i+1)$ belongs to $\text{Par}_2((v, i))$ and the proof is complete.

Now, assume that $d_G(v, s) \leq 2^{i+2} - 2$, and set p denotes the shortest path in G that connects graph vertex v to source vertex s. Assume that path p is partitioned to two subpaths p_1 and p_2 such that p_1 is a path of length $2^{i+1} - 1$ from v to x, and p_2 is a path from x to s. Since $p = p_1 \cup p_2$ and $|p_1| = 2^{i+1} - 1$, we deduce that:

$$|p_2| = d_G(v, s) - 2^{i+1} - 1 \tag{2.13}$$

Since $d_G(v, s)$ is greater than $2^{i+2} - 2$, the value of $|p_2|$ is positive. This implies that it is possible to partition p in this way. The following figure depicts vertices s, v, x, and the paths between them.

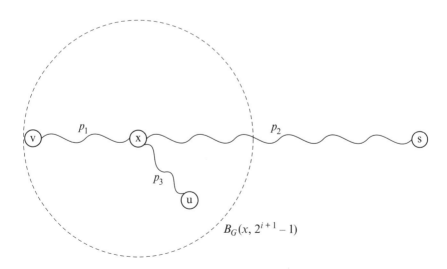

Now, we continue the proof in two possible cases: $x \in I_{i+1}$ and $x \notin I_{i+1}$. In the first case, since $d_G(x, v) \leq |p_2| < 2^{i+2} - 2$, x is a member of set $I_{i+1} \cap B_G(v, 2^{i+2} - 2)$. Henceforth, according to equation (2.12), $\text{Par}_2((v, i)) \neq \varnothing$.

In the latter case, we show that there is some vertex u in I_{i+1} such that $u \in B_G(v, 2^{i+1} - 1)$. By contradiction, assume that there is no such vertex. Hence, we obtain that:

$$\forall u \in I_{i+1} : d_G(x, u) \geq 2^{i+1}$$

which means that x is so far from all of the I_{i+1} members that by adding x to set I_{i+1}, it still remains a 2^{i+1}-independent set. In addition, since x does not belong to I_{i+1}, $|I_{i+1} \cup \{x\}| > |I_{i+1}|$ which contradicts the assumption that I_{i+1} is a *maximal* 2^{i+1}-independent set. Henceforth, we have shown that:

$$\exists u \in I_{i+1} : u \in B_G(x, 2^{i+1} - 1)$$

Let p_3 denote the shortest path in G from x to u. Here, we find an upper bound of $d_G(u, v)$.

$$d_G(u, v) \leq d_G(u, x) + d_G(x, v) \leq |p_1| + |p_2|$$

Since $|p_1| = 2^{i+1} - 1$, we obtain the following inequality:

$$d_G(u, v) \leq (2^{i+1} - 1) + (2^{i+1} - 1) = 2^{i+2} - 2 \tag{2.14}$$

Regarding inequality (2.14) and the definition of Part$_2$, we conclude that Par$_2((v, i)) \neq \emptyset$.

The rest of the proof is the same as what we did to prove that HIT type-1 is connected and acyclic. ∎

After presenting the existence proof of the HIT type-2, we will see some of its properties.

Lemma 2.6 Let $T^s = (V_{T^s}, E_{T^s})$ be a hierarchical independence tree type-2 of HIS $\bar{I}_s = (I_0, I_1, \ldots, I_h)$ in graph $G = (V, E)$.

1. $\forall v \in I_i, \forall u \in \mathcal{L}_{T^s}(v, i) : d_G(u, v) < 2^{i+2}$ $\quad \forall i = 0, 1, \ldots, h$
2. $\forall v \in I_i, \forall u_1, u_2 \in \mathcal{L}_{T^s}(v, i) : d_G(u_1, u_2) < 2^{i+3}$ $\quad \forall i = 0, 1, \ldots, h$
3. $B_G(s, 2^i - 1) \subseteq \mathcal{L}_{T^s}(s, i)$ $\quad \forall i = 0, 1, \ldots, h$

where $\mathcal{L}_{T^s}(v, i) = \{u : (u, 0) \in T^s_{(v,i)} \cdot \text{leaves}\}$.

Proof. The proof of the first property is similar to what you previously saw in the proof of Lemma 2.4 for the HIT type-1. The only difference is that for the HIT type-2:

$$|p_c| \leq 2^{i+2} - 2 < 2^{i+2}$$

Subsequently,

$$|p| = \sum_{c=0}^{i-1} 2^{c+2} < 2^{i+2}$$

Since $d_G(u, v) \leq |p|$, the proof is complete.

For the second property, we use the following inequality:

$$\forall u_1, u_2 \in \mathcal{L}_{T^s}(v, i) : d_G(u_1, u_2) \leq d_G(u_1, v) + d_G(v, u_2)$$

Consequently, according to the first property, we obtain the following inequality:

$$d_G(u_1, u_2) < 2^{i+2} + 2^{i+2} \leq 2^{i+3}$$

To prove the last property, we will show that for every $j = 0, 1, \ldots, i$, if v belongs to $I_j \cap B_G(s, 2^i - 1)$, ordered pair (v, i) is a vertex in $T^s_{(s,i)}$. We do this by induction on j.

Let v denote an arbitrary vertex in $I_i \cap B_G(s, 2^i - 1)$. If $v \neq s$, since v is in set I_i, $d_G(v, s)$ will not be less than 2^i, which contradicts the assumption that $v \in B_G(s, 2^i - 1)$. As a result, $v = s$, and $(v, i) = (s, i)$ is a vertex in tree $T^s_{(s,i)}$ (the first base case of the induction).

Additionally, assume that $v \in I_{i-1} \cap B_G(s, 2^i - 1)$. Regarding the definition of Par_2, $\mathrm{Par}_2((v, i - 1)) = \{(s, i)\}$, because $s \in I_i$, $s \in B_G(s, 2^{i+1} - 2)$, and s is the closest vertex to itself that belongs to set $I_i \cap B_G(v, 2^{i+1} - 2)$. As (s, i) is the only member of $\mathrm{Par}_2((v, i - 1))$, it is chosen as the parent of $(v, i - 1)$. Consequently, $(v, i - 1)$ is a vertex in subtree $T^s_{(s,i)}$ (the second base case of the induction).

Now, assume that for some $j = 1, 2, \ldots, i - 1$, if $y \in I_j \cap B_G(s, 2^i - 1)$, (y, j) will be a vertex of $T^s_{(s,i)}$ (Induction Hypothesis). Additionally, consider that $v \in I_{j-1} \cap B_G(s, 2^i - 1)$. Here, we continue the proof by considering two possible cases: $d_G(v, s) \leq 2^{j+1} - 2$ and $d_G(v, s) > 2^{j+1} - 2$. In the earlier one, as vertex s belongs to $I_j \cap B_G(v, 2^{j+1} - 2)$, it is the closest vertex to itself that holds this condition, i.e., $\mathrm{Par}_2((v, j - 1)) = \{(s, i)\}$. Subsequently, tree vertex (s, j) is the parent of $(v, j - 1)$. On the other hand, regarding the induction hypothesis, since $s \in I_j \cap B_G(s, 2^i - 1)$, vertex (s, j) is a vertex in subtree $T^s_{(s,i)}$. Consequently, $(v, j - 1)$ is also a vertex in $T^s_{(s,i)}$.

In the latter case, let p denote the shortest path from v to s in G. We divide p into two subpaths p_1 and p_2 such that p_1 is a path from v to x such that $|p_1| = 2^j - 1$; consequently, p_2 is a path from x to s of length $|p_1| = 2^j - 1$. Now, there are two possibilities: either $x \notin I_j$ or $x \in I_j$.

1. $x \notin I_j$: Assume that there is no vertex $u \in B_G(x, 2^j - 1)$ such that $u \in I_j$. Henceforth, for every $g \in I_j : d_G(g, x) \geq 2^j$. This implies that $I_j \cup \{x\}$ is a 2^j-independent set. As far as $x \notin I_j$, $|I_j \cup \{x\}|$ is greater than $|I_j|$ which contradicts the assumption that I_j is a maximal 2^j-independent set. Subsequently, there is some graph vertex $u \in B_G(x, 2^j - 1) \cap I_j$. Let p_3 denote the shortest path from u to x in G ($|p_3| \leq 2^j - 1$). This implies that $p_1 \cup p_3$ is a path from u to v of length $|p_1| + |p_3|$ which is not greater than $(2^j - 1) + (2^j - 1) = 2^{j+1} - 2$. Hence, $u \in B_G(v, 2^{j+1} - 2) \cap I_j$. Moreover, $p_2 \cup p_3$ is a path from u to s of the following length $|p_2| + |p_3|$. Consider the following inequalities:

$$|p_2|+|p_3| \le (|p|-2^j+1)+(2^j-1) \le |p| \le 2^i-1$$

This implies that $d_G(u, s) \le 2^i - 1$. Regarding the definition of Par_2, since, $u \in B_G(v, 2^{j+1} - 2) \cap I_j$, and $d_G(u, s) \le 2^i - 1$, we conclude that:

$$\forall q \in \text{Par}_2((v, j-1)): d_G(q, s) \le d(u, s) \le 2^i - 1$$

As a result, the parent of $(v, j - 1)$ belongs to $B_G(s, 2^i - 1)$. Consequently, according to the induction hypothesis, since the parent of $(v, j - 1)$ is a vertex in tree $T^s_{(s,i)}$, vertex $(v, j - 1)$ also belongs to the vertex set of $T^s_{(s,i)}$.

2. $x \in I_j$: As far as $d_G(v, x) \le |p_1| = 2^j - 1$, vertex x satisfies the following condition:

$$x \in I_j \cap B_G(v, 2^{j+1} - 2) \tag{2.15}$$

Here, there are two possible cases. In the first one, x is the closest vertex to s that satisfies the above condition. Concerning the definition of Par_2, $(x, j) \in \text{Par}_2((v, j - 1))$, and for every $(h, j) \in \text{Par}_2((v, j - 1))$, the following inequalities hold:

$$d_G(h, s) \le d_G(x, s) \le 2^j - 1 \le 2^i - 1$$

As a result, we obtain the following proposition:

$$\forall (q, j) \in \text{Par}_2((v, j-1)): d_G(q, s) \le 2^i - 1 \tag{2.16}$$

Hence, if (z, j) denotes the parent of $(v, j - 1)$ in tree T^s, $d_G(z, s)$ is not greater than $2^i - 1$, which implies that $z \in I_j \cap B_G(s, 2^i - 1)$. Consequently, regarding the induction hypothesis, (z, j) is a vertex in $T^s_{(s,i)}$; as a result, $(v, j - 1)$ also belongs to the vertex set of $T^s_{(s,i)}$.

In the latter case, x is not the closest vertex to s that satisfies condition (2.15). Again, according to the definition of Par_2, inequality (2.16) holds. Using the same reasoning as we did for the first case, we conclude that $(v, j - 1)$ is a vertex of $T^s_{(s,i)}$.

Up to now, we have shown that $\forall j = 0, 1, \ldots, h$, if $v \in I_j \cap B_G(s, 2^i - 1)$, (v, j) will belong to the vertex set of $T^s_{(s,i)}$. In the case that $j = 0$, taking into consideration that $I_0 = V$, we conclude that:

$$v \in B_G(s, 2^i - 1) \rightarrow (v, 0) \text{ is a vertex in } T^s_{(s,i)}$$
$$\rightarrow v \in \mathcal{L}_{T^s}(s, i)$$

Consequently, $B_G(s, 2^i - 1) \subseteq \mathcal{L}_{T^s}(s, i)$, and the proof is complete. ∎

2.5　Hierarchical Independence Tree Type-0

In the previous sections, we have become familiar with two types of hierarchical independence trees. They have some common properties and also some differences. In this section, we introduce the third type of HIT, which is more efficient than the previous ones when considering the versatile routing schemes that will be presented in the next chapter.

Previously, the hierarchical independent sequence of some source vertex in a connected graph was defined. Here, we define a special class of the HIS that will be used in defining the HIT type-0.

Definition 2.6   Let $G = (V, E)$ denote a connected unweighted graph. Assuming distinguished vertex $s \in V$ as the source vertex, sequence $\bar{I}_s = (I_0, I_1, \ldots, I_h)$ is the *source-oriented hierarchical independent sequence* (source-oriented HIS), if both of the following hold:

- $I_h = \{s\}$.
- For every $i = 0, 1, \ldots, h - 1$ and $r \geq 2^i$, $I_i \cap A_r$ is a maximal 2^i-independent set of subgraph G_{A_r} where $A_r = B_G(s, r)$.

Note that we will refer to the previous version of the hierarchical independence sequence as the *basic* HIS. In Definition 2.6, set I_{i+1} is not necessarily a subset of I_i (for every $i < h - 1$). This implies that every source-oriented HIS is not necessarily a basic one.

In the next chapter, it will be shown that for every connected graph and source vertex, there exists a source-oriented HIS; however, like the basic HIS, there may exist more than one source-oriented HIS for a given graph and source vertex.

The HIT type-0 is different from that of type-1 in two ways: it is defined based on the source-oriented HIS and it uses the parent function type-0 defined by the following equation.

$$\text{Par}_0((v, i)) = \{(u, i+1) : u \in I_{i+1} \cap B_G(v, 2^{i+1} - 1), d_G(u, s) \leq d_G(v, s)\}$$
$$\forall v \in I_i, \forall i = 0, 1, \ldots, h - 2 \tag{2.17}$$

Note that in the definition of HIT type-0, instead of forcing the parent vertex be the closest one to the source, it only needs *to not be further* than its child from source vertex s.

Let T^s be an HIT type-0 of source-oriented HIS $\bar{I}_s = (I_0, I_1, \ldots, I_h)$ in graph $G = (V, E)$. If v denotes an arbitrary member of I_i, we obtain the following:

1. For every vertex $v \in I_{i+1} \cap I_i$, $(v, i - 1) \in \text{Par}_0((v, i))$.
2. For every vertex $u \in \mathcal{L}_{T^s}(v, i)$, $d_G(u, v) < 2^{i+1}$.
3. For every pair of vertices $u_1, u_2 \in \mathcal{L}_{T^s}(v, i)$, $d_G(u_1, u_2) < 2^{i+2}$.
4. For every $i = 0, 1, \ldots, h$, $B_G(s, 2^i - 1) \subseteq \mathcal{L}_{T^s}(s, i)$.

For all of these properties, $\mathcal{L}_{T^s}(v, i) = \{l : (l, 0) \in T^s_{(v,i)} \cdot \text{leaves}\}$. The first three properties of the HIT type-0 are held in common with the HIT type-1 and can be proved in the same way. Additionally, the last property is in common with the HIT type-2 and will be proved in Lemma 2.8.

The following lemma guarantees the existence of the HIT type-0 for any source-oriented HIS.

Lemma 2.7 For any source-oriented HIS $\bar{I}_s = (I_0, I_1, \ldots, I_h)$ of source vertex s in graph $G = (V, E)$, there is a hierarchical independence tree type-0.

Proof. Like the HIT type-2, we only need to show that:

$$\forall v \in I_i : \mathrm{Par}_0((v, i)) \neq \varnothing$$

We consider the following three cases:

1. $v \in I_{i+1}$.
2. $v \in B_G(s, 2^{i+1} - 1)$.
3. $v \in I_{i+1} \wedge v \notin B_G(s, 2^{i+1} - 1)$.

If $v \in I_{i+1}$, regarding the first property of the HIT type-0, $(v, i + 1) \in \mathrm{Par}_0(v, i)$. Moreover, in the case that $v \in B_G(s, 2^{i+1} - 1)$, according to the definition of function Par_0, $s \in \mathrm{Par}_0((v, i))$.

Now consider the case that $v \in I_{i+1}$ and $v \in B_G(s, 2^{i+1} - 1)$. If we prove the following proposition, we will conclude that $u \in \mathrm{Par}_0((v, i))$ and the proof will be complete.

$$\exists u \in I_{i+1} : u \in B_G(v, 2^{i+1} - 1) \cap B_G(s, d_G(v, s)) \qquad (2.18)$$

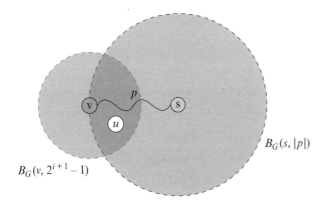

$B_G(v, 2^{i+1} - 1)$

$B_G(s, |p|)$

By contradiction, assume that proposition (2.18) is not true. Subsequently,

$$\forall u \in I_{i+1} : u \notin B_G(v, 2^{i+1} - 1) \cap B_G(s, d_G(v, s))$$

or equivalently,

$$\forall u \in I_{i+1} \cap B_G(s, d_G(v, s)) : u \notin B_G(v, 2^{i+1} - 1) \qquad (2.19)$$

In addition, according to Definition 2.6, $I_{i+1} \cap B_G(s, d_G(v, s))$ is a maximal 2^{i+1}-independent set in the subgraph of G induced by $B_G(s, d_G(v, s))$. We will refer to this subgraph as G'.

Proposition (2.19) implies that all of the members of I_{i+1} in G' are so far from vertex v that if we add v to set $I_{i+1} \cap B_G(s, d_G(v, s))$, the resulting set will remain a 2^{i+1}-independent set in G'. Since vertex v does not belong to $I_{i+1} \cap B_G(s, d_G(v, s))$, it is inferred that:

$$|\{v\} \cup I_{i+1} \cap B_G(s, d_G(v, s))| = |I_{i+1} \cap B_G(s, d_G(v, s))| + 1 > |I_{i+1} \cap B_G(s, d_G(v, s))|$$

which contradicts our previous inference that $I_{i+1} \cap B_G(s, d_G(v, s))$ is a maximal 2^{i+1}-independent set in G'. ■

Finally, we show an important property of the HIT type-0, which is in common with the HIT type-2.

Lemma 2.8 Let T^s denote an HIT type-0 of source-oriented HIS $\overline{I}_s = (I_0, I_1, \ldots, I_h)$ in graph $G = (V, E)$. For every $i = 0, 1, \ldots, h$, the following relation holds:

$$B_G(s, 2^i - 1) \subseteq \mathcal{L}_{T^s}(s, i)$$

Proof. To prove this property, like the proof of Lemma 2.6, we first use the induction on j to show that:

$$\forall j = 0, 1, \ldots, i \forall v \in I_j \cap B_G(s, 2^i - 1) : (v, j) \text{ is a vertex in subtree } T^s_{(s,i)} \qquad (2.20)$$

As the base case, since $I_i \cap B_G(s, 2^i - 1) = \{s\}$, and (s, i) is the root of $T^s_{(s,i)}$, we are done (note that for every vertex q in $I_i \cap B_G(s, 2^i - 1)$, since $d_G(s, q) \leq 2^i - 1$, graph vertices q and s cannot be in the same 2^i-independent set (I_i) simultaneously unless $q = s$).

Now, assume that for some $k = 1, 2, \ldots, i$, for every vertex $y \in I_k \cap B_G(s, 2^i - 1)$, (y, k) will be a vertex in $T^s_{(s,i)}$ (induction hypothesis). Additionally, consider the arbitrary vertex v such that:

$$v \in I_{k-1} \cap B_G(s, 2^i - 1)$$

As vertex v belongs to I_{k-1}, there exists a tree vertex in the form $(v, k - 1)$. According to the definition of function Par_0, if (u, k) denotes the parent of $(v, k - 1)$ in tree T^s, we infer that:

$$(u, k) \in \text{Par}_0((v, k-1)) \rightarrow u \in I_k \wedge u \in B_G(v, 2^k - 1) \wedge d_G(u, s) \leq d_G(v, s)$$

In addition, since vertex $u \in B_G(s, 2^i - 1)$, the following inequality holds:

$$d_G(v, s) \leq 2^i - 1$$

Moreover, as far as $d_G(u, s) \leq d_G(v, s)$, we obtain that:

$$d_G(u, s) \leq 2^i - 1 \rightarrow u \in B_G(s, 2^i - 1)$$

This implies that $u \in I_k \cap B_G(s, 2^i - 1)$. As a result, regarding the induction hypothesis, (u, k) is a vertex in subtree $T_{(s,i)}^s$. Since (u, k) is the parent of $(v, k-1)$, vertex $(v, k-1)$ also belongs to the vertex set of subtree $T_{(s,i)}^s$.

Up to now, we have proved proposition (2.20) for every $j \leq i$. Considering the case that $j = 0$, we obtain the following proposition:

$$\forall v \in I_0 \cap B_G(s, 2^i - 1) : (v, 0) \text{ is a vertex in subtree } T_{(s,i)}^s$$

or equivalently ($I_0 = V$),

$$\forall v \in B_G(s, 2^i - 1) : (v, 0) \text{ is a vertex in subtree } T_{(s,i)}^s$$

which implies that:

$$B_G(s, 2^i - 1) \subseteq \mathcal{L}_{T^s}(s, i) \qquad \blacksquare$$

2.6   Summary and Outlook

In this chapter, we presented four different hierarchical data structures that will be used in the next chapter as the routing tools to solve the oblivious routing problems in a versatile approach.

In the first section, we presented some preliminary definitions to use them in the rest of the chapter. In the second one, we introduced the HDT, which is a top-down hierarchical tool to organize the routing-related data in a connected graph.

In the next three sections, we addressed three different types of the HIT as an example of the bottom-up hierarchical routing tool. The properties of each type were presented and proved mathematically.

A brief comparison of the three types of HITs and the HDT presented in this chapter is given in table 2.1.

Exercises

1. Prove the properties of a graph Δ-partition enumerated in section 2.1.
2. Extend Definition 2.2 in the following way: for some positive value Δ, we call set $X \subseteq 2^V$ a *maximal Δ-partition* of graph G if the following two conditions hold regarding X:
   a. X is a Δ-partition of graph G.
   b. For every $\delta \in (0, \Delta)$, X is *not* a δ-partition of G.
   Now, assume that X_Δ represents an arbitrary maximal Δ-partition of graph G (for every $\Delta > 0$). By decreasing the value of Δ, the maximum possible

Table 2.1 Comparison of HDT, HIT Type-0, HIT Type-1, and HIT Type-2

Tree	Problem type	Approach	Hierarchical sequence	Graph	Partition	Child-parent distance	Parent function
HDT[a]	Multiple source	Top-down	HDS	Connected weighted	2^i-Partition	N/A	N/A
HIT type-0[b]	Single source	Bottom-up	Source-Oriented HIS	Connected unweighted	2^{i+2}-Partition	$\leq 2^{i+1} - 1$	Par_0
HIT type-1	Single source	Bottom-up	Basic HIS	Connected unweighted	2^{i+2}-Partition	$\leq 2^{i+1} - 1$	Par_1
HIT type-2[c]	Single source	Bottom-up	Basic HIS	Connected unweighted	2^{i+3}-Partition	$\leq 2^{i+2} - 2$	Par_2

[a]Presented by Gupta et al. (2006).
[b]Our tree.
[c]Presented by Srinivasogopalan, Busch, and Iyengar (2012).

distance between the vertices of the clusters in X_Δ will also decrease. Subsequently, it seems that the cardinality of each cluster decreases as well. Intuitively, this implies that for any pair of positive values Δ_1 and Δ_2, if $\Delta_1 < \Delta_2$, X_{Δ_1} should have more members (than X_{Δ_2}) to cover all the graph vertices. However, this claim is not generally true. Why? (Hint: Use the graph example shown in 2.1 to contradict the claim.)

3. Assume that the *independence set I* of graph $G = (V, E, w)$ is defined as the set of vertices in G for which its induced graph G_I has no edges. What is the relation between an independent set of G and an r-independent set of graph G? Explain your reasoning.

4. Prove the basic properties of the HIS of a graph enumerated in section 2.3.

5. Briefly compare the hierarchical data structures presented in this chapter. Moreover, explain how well they can be used as a tool for versatile routing. What are the pros and cons of each one?

6. Considering Definition 2.5, prove that sequence $(\mathcal{V}_0, \mathcal{V}_0, \mathcal{V}_0, \mathcal{V}_1, \mathcal{V}_2, \dots, \mathcal{V}_{h-3}, \{V\})$ specifies an HDS of graph G.

7. Prove the first three properties of the HIT type-0 mentioned in section 2.5.

Suggested Reading

Bartal, Y. 1996. Probabilistic approximation of metric spaces and its algorithmic applications. *Proceedings of the 37th FOCS*, 184–193.

Garg, N., R. Khandekar, G. Konjevod, R. Ravi, F. S. Salman, and A. Sinha. 2001. On the integrality gap of a natural formulation of the single-sink buy-at-bulk network design formulation. *Proceedings of the 8th IPCO*, 170–184.

Guha, S., A. Meyerson, and K. Munagala. 2000. Hierarchical placement and network design problems. *Proceedings of the 41st FOCS*, 603–612.

Gupta, A., M. T. Hajiaghayi, and H. Racke. 2006. Oblivious network design. *Proceedings of the Seventeenth Annual ACM-SIAM Symposium on Discrete Algorithms, SODA '06*. New York: ACM, 970–979.

Harrelson, C., K. Hildrum, and S. B. Rao. 2003. A polynomial-time tree decomposition to minimize congestion. *Proceedings of the 15th SPAA*, 34–43.

Kleinberg, J., A. Slivkins, and T. Wexler. 2009. Triangulation and embedding using small sets of beacons. *Journal of the ACM*, 56(6):1–37.

Kuhn, F., T. Moscibroda, and R. Wattenhofer. 2005. On the locality of bounded growth. *Proceedings of PODC '05*. New York: ACM, 60–68.

Nieberg, T. 2006. Independent and dominating sets in wireless communication graphs. Doctoral dissertation, University of Twente, Zwolle.

Salman, F. S., J. Cheriyan, R. Ravi, and S. Subramanian. 2000. Approximating the single-sink link-installation problem in network design. *SIAM Journal on Optimization*, 11(3):595–610.

Srinivasagopalan, S. 2011. Oblivious buy-at-bulk network design algorithms. Doctoral dissertation, Louisiana State University, Baton Rouge, LA.

Srinivasagopalan, S., C. Busch, and S. S. Iyengar. 2009. Brief announcement: Universal data aggregation trees for sensor networks in low doubling metrics in algorithmic aspects of wireless sensor networks. *Proceedings of the 5th International Workshop, ALGOSENSORS*. Berlin: Springer-Verlag, 151–152.

Srinivasagopalan, S., C. Busch, and S. S. Iyengar. 2011a. An oblivious spanning tree for single-sink buy-at-bulk in low doubling-dimension graphs. Louisiana State University, Baton Rouge, LA.

Srinivasagopalan, S., C. Busch, and S. S. Iyengar. 2011b. Oblivious buy-at-bulk in planar graphs. *Workshop on Algorithmic Computing, WALCOM*, IIT-Delhi, 18–20 February.

Chapter 3   Routing Schemes in Oblivious Network Design

We are constantly overwhelmed by random events. This is because of our lack of knowledge about many surrounding facts. Some random events are very hard to predict, such as the rain, thunderstorms, high traffic volume, and the failure of a data connection in an Internet data link.

Networks also work as connecting tools in the context of such uncertain environments. With this in mind, it is vital to find adaptable and versatile approaches to route network flows in non-deterministic conditions. As mentioned in chapter 1, this uncertainty can be modeled using an *oblivious* routing environment, which considers a wide range of possible cases. Additionally, we suggested two different approaches to deal with the oblivious routing environment: the dynamic and versatile approach. As far as large-sized networks are concerned, the versatile approach is much more suitable for implementation.

The proliferation of network routing schemes has promoted massive network design to achieve low latency and high reliability with optimal cost. In this chapter, we will introduce two algorithms that use versatile routing schemes to flexibly solve large-sized oblivious routing problems. More specifically, we will address two different versatile routing schemes to solve oblivious routing problems efficiently. The first scheme is based on the top-down hierarchical routing tree mentioned in the previous chapter. However, the second scheme uses HITs, which organize data in a bottom-up fashion. We also analyze the routing cost incurred by each of the routing schemes by computing their competitive ratios (see section 1.3).

3.1   A Top-Down Versatile Routing Scheme

In this section, the versatile routing scheme presented by Gupta, Hajiaghayi, and Racke (2006) will be described and analyzed thoroughly.

This scheme benefits from the use of the top-down hierarchical decomposition tree described in chapter 2.

In all of the discussions made in this section regarding weighted connected graph $G = (V, E, w)$, we assume (without loss of generality) that the minimum distance between any two distinct vertices of a graph is greater than one, i.e.:

$$w(e) > 1 \quad \forall e \in E \tag{3.1}$$

Additionally, the graph diameter is of the following form:

$$\text{diam}(G) = 2^h \quad \text{for some } h \in \mathbb{N} \tag{3.2}$$

The following lemma shows that these two assumptions do not restrict our discussion about weighted connected graphs to a special case.

Lemma 3.1 For any connected graph $G = (V, E, w)$, there is a connected graph $G_c = (V, E, w_c)$ such that for every edge $e \in E$, $w_c(e) = c \cdot w(e)$ $(c \in \mathbb{R}^+)$, $\text{diam}(G_c) = 2^h$ (for some $h \in \mathbb{Z}^+$), and for any two different vertices $u, v \in V$, $d_{G_c}(u, v) > 1$.

Proof. Consider graph $G' = (V, E, w')$ where:

$$w'(e) = \frac{1 + \varepsilon}{min_{e \in E} w(e)} \cdot w(e) \quad \forall e \in E$$

such that $\varepsilon \in \mathbb{R}^+$ is some small positive number. It is easy to prove that $\forall u, v \in V$, if $u \neq v$, $d_{G'}(u, v) > 1$. Now, we define graph $G'' = (V, E, w'')$ such that:

$$w''(e) = \frac{2^{\lceil \log_2 \text{diam}(G') \rceil}}{\text{diam}(G')} \cdot w'(e) \quad \forall e \in E$$

Note that concerning the above definition of w'', $\text{diam}(G'')$ is 2^h, for h = $\lceil \log_2 \text{diam}(G') \rceil$. Moreover, since

$$\frac{2^{\lceil \log_2 \text{diam}(G') \rceil}}{\text{diam}(G')} \geq 1$$

$w''(e) \geq w'(e)$ for every $e \in E$. As a result, for the following value of c, graph $G_c = (V, E, w_c)$ satisfies the aforementioned conditions.

$$c = \frac{1 + \varepsilon}{min_{e \in E} w(e)} \cdot \frac{2^{\lceil \log_2 \text{diam}(G') \rceil}}{\text{diam}(G')} \qquad \blacksquare$$

In the remainder of this section, we first specify the most general case of the oblivious routing problem, which can be solved using Gupta's routing scheme. Then, we introduce a special class of the hierarchical decomposition sequence known as *padded* HDS. Additionally, a randomized algorithm will be presented to generate a padded HDS. Finally, using the padded HDS of a graph, a versatile routing scheme will be introduced and its competitive ratio will be computed.

3.1.1 Routing Problem Specification

In chapter 1, we introduced a specific type of the general routing problem in which the routing cost environment is oblivious. In other words, in such problems, instead of specifying one routing cost environment, we have a set of possible routing cost environments ($\mathbb{E}$), which makes the routing cost of the problem non-deterministic.

In this section, we are focusing on the oblivious routing problem of graph $G = (V, E, w)$ which satisfies conditions (3.1) and (3.2). Moreover, the set of possible routing cost environments of the problem is in the following form:

$$\mathbb{E} = \left\{ (\text{cost}_e, \Sigma, \bar{K}): \text{cost}_e \in F_{\text{sub additive}}, \bar{K} \in \mathcal{K} \right\} \qquad (3.3)$$

such that Fsub-additive denotes the set of all sub-additive[1] functions in the following form:

$$\text{cost}_e(f_1, f_2, \dots, f_k) = w(e) \cdot \text{rrc}(f_1, f_2, \dots, f_k) \quad \text{where} \quad \text{rrc}: \mathbb{R}_{\geq 0}^k \mapsto \mathbb{R}_{\geq 0}$$

and where $\mathcal{K}$ represents the set of all the possible sequences of k commodities[2] in graph G ($\forall k \in \mathbb{N}$).

3.1.2 Padded Hierarchical Decomposition Sequence

In the previous chapter, the hierarchical decomposition sequence of a connected graph was introduced. Now, we define a special case of the HDSs in which each vertex is far away from the boundaries of its containing cluster at every level of partitioning. This padding property is important for the routing cost analysis of the versatile routing scheme, which will be presented later in this chapter.

Definition 3.1 Consider positive number $\alpha \leq 1$ and connected graph $G = (V, E, w)$ of diameter 2^h. For hierarchical decomposition sequence $\bar{H} = (H_0, H_1, \dots, H_h)$ of graph G, vertex v is α-padded in $\bar{H}$ if:

$$\forall i = 0, 1, \dots, h : \exists C \in H_i \text{ such that } B_G(v, \alpha \cdot 2^i) \subseteq C \qquad (3.4)$$

---

1. Assuming that A and B are two subsets of real numbers set $\mathbb{R}$, function $f: A^k \mapsto B$ is sub-additive if for every $x_1, y_1, x_2, y_2, \dots, x_k, y_k \in A$, we obtain that: $f(x_1 + y_1, x_2 + y_2, \dots, x_k + y_k) \leq f(x_1, x_2, \dots, x_k) + f(y_1, y_2, \dots, y_k)$.

2. Regarding the definition of commodity in section 1.2, every commodity is a triple of source vertex, target vertex, and commodity value.

An illustration of the above definition is given in figure 3.1, which schematically depicts a padded vertex in an HDS of a graph of diameter 8 ($h =$ 3). In this figure, every cluster is represented by a gray circle such that the shown parameter (D) denotes the maximum distance between the cluster members. Note that partition H_3 has only one cluster, as represented by the largest circle in figure 3.1. The only cluster is then partitioned into three smaller clusters that are included in H_2. The upper cluster of H_2 is also partitioned into three smaller clusters that are members of H_1. Finally, the basic cluster containing vertex v has been determined (according to our assumption that the distance between any two distinct vertices is greater than one, every basic cluster contains *one and only one* vertex). We see that

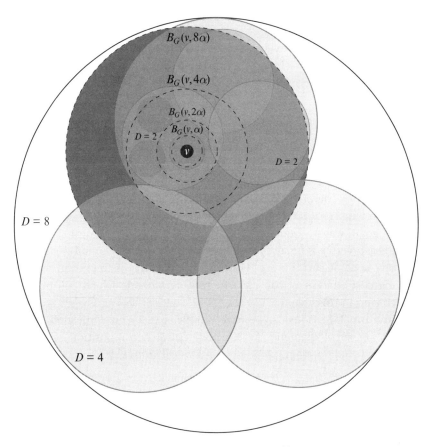

Figure 3.1 Schematic view of a padded vertex in HDS $\bar{H} = (H_0, H_1, H_2, H_3)$ of some graph.

proposition (3.4) is true for the shown vertex (v); henceforth, vertex v is α-padded in HDS $\bar{H}$.

Fakcharoenphol's Algorithm In 2003, Fakcharoenphol et al. presented a randomized algorithm to generate a random HDS in which a quotient of vertices are α-padded with high probability ($\alpha \leq 1/8$). The main idea of this algorithm is to make a partition in the ith level by putting some portion of the vertices of each upper-level cluster into a new smaller one such that the members of the new cluster would not be farther than X_i from a special vertex called the *representative* vertex of the cluster, where X_i is a uniformly distributed random variable in the interval $[2^i/4, 2^i/2)$. A more detailed description is given in algorithm 3.1.

In this algorithm, a connected weighted graph is given as the input of the algorithm and a random HDS of the graph will result as the output. Additionally, every vertex v has three attributes: "*rep*" which is the *representative* vertex of v, "*cluster*," which is a set of vertices having a single representative v, and the Boolean "*flag*."

In the first two lines of this algorithm, we generate the uniformly random permutation π on vertex set V and random number $\mathcal{U}$, which is uniformly distributed over interval $[1/2, 1)$. Then, we start making the sequence of partitions by assigning $\{V\}$ to H_h. Consider the loop expanded from line 4 to 25, which makes other partitions. To obtain partition H_i from H_{i+1}, we first specify a representative for each vertex $v \in V$ in the following way: assuming $C \in H_{i+1}$ as the only H_{i+1} member that contains v, the representative of v is the first vertex in permutation π such that $v \in C \cap B_G(v, \mathcal{U} \cdot 2^{i-1})$ (see line 9). After specifying the representatives of all the vertices, attribute *cluster* of vertex v is developed as the set of vertices whose representatives are equal to v (see lines 11 to 18). Finally, partition H_i is obtained by gathering all of the nonempty developed clusters together into a set (lines 19–23).

Algorithm 3.1 constructs random HDS $\bar{H}$ that satisfies the following condition (for $\alpha \leq 1/8$).

$$\mathbf{Pr}\left[v \text{ is not } \alpha\text{-padded in } \bar{H}\right] \leq O(\alpha \log |V|) \quad \forall v \in V \tag{3.5}$$

This inequality guarantees that each graph vertex is α-padded with high probability in the random HDS generated by algorithm 3.1. To prove this claim, we first show that the resulting partition sequence of algorithm 3.1 is an HDS; then, we prove inequality (3.5).

Lemma 3.2 Algorithm 3.1 constructs an HDS in the input graph.

Proof. Regarding the third line of the algorithm, $H_h = \{V\}$. Consider partition H_i for every $i \leq h - 1$. Since all the members of a cluster in H_i have the

---

Algorithm 3.1   Randomized HDS generator

Input: Connected graph $G = (V, E, w)$ of diameter 2^h

Output: HDS $\bar{H} = (H_0, H_1, \ldots, H_h)$ of graph G

1    $\pi \leftarrow$ A uniformly random permutation on members of V;

2    $\mathcal{U} \leftarrow \mathrm{unif}[1/2, 1)^1$

3    $H_h \leftarrow \{V\}$;

4   **for** $i \leftarrow h - 1$ **to** 0 **do**

5       $H_i \leftarrow \varnothing$;

6   **foreach** $C \in H_{i+1}$ **do**

7   **foreach** $v \in C$ **do**

8           $v.\mathrm{cluster} \leftarrow \varnothing$; $v.\mathrm{flag} \leftarrow$ **true**;

9           $v.\mathrm{rep} \leftarrow \pi\left(\min_j\{\pi(j) \in C \cap B_G(v, \mathcal{U}.2^{i-1})\}\right)$;

10   **end**

11   **foreach** $v \in C$ **do**

12   **foreach** $u \in C$ **and** $u.\mathit{flag} =$ **true** **do**

13   **if** $u.\mathrm{rep} = v$ **then**

14             $u.\mathit{flag} \leftarrow$ **false**;

15             $v.\mathrm{cluster} \leftarrow v.\mathrm{cluster} \cup \{u\}$;

16   **end**

17   **end**

18   **end**

19   **foreach** $v \in C$ **do**

20   **if** $v.\mathrm{cluster} \neq \varnothing$ **then**

21           $H_i \leftarrow H_i \cup \{v.\mathrm{cluster}\}$;

22   **end**

23   **end**

24   **end**

25   **end**

---

[1]Uniform random variable in interval $\left[\dfrac{1}{2}, 1\right)$

same representative vertex, for every two different vertices v_1 and v_2 in a cluster, $u = v_1 \cdot \text{rep}$ and $u = v_2 \cdot \text{rep}$. Henceforth, concerning line 9 of algorithm 3.1, we obtain the following relations:

$$u \in B_G\left(v_1, \mathcal{U} \cdot 2^{i-1}\right) \rightarrow d_G(v_1, u) \le \mathcal{U} \cdot 2^{i-1}$$

and

$$u \in B_G\left(v_2, \mathcal{U} \cdot 2^{i-1}\right) \rightarrow d_G(v_2, u) \le \mathcal{U} \cdot 2^{i-1}$$

Subsequently, regarding the triangle inequality, we infer that ($\mathcal{U} < 1$):

$$d_G(v_1, v_2) \le d_G(v_1, u) + d_G(v_2, u) \le 2^{i-1} + 2^{i-1} \le 2^i$$

Up to this point, we have shown that for every $i < h$, H_i is a 2^i-partition of the input graph. Additionally, we need to prove that for every $i < h$, assuming that cluster C belongs to H_i, there exists cluster $C' \in H_{i+1}$ such that $C \subseteq C'$. Since every iteration of the for statement expanded from line 6 to 24 uses one distinct member of H_{i+1} to construct a cluster of H_i, the mentioned condition holds for every $i < h$. ■

In order to show inequality (3.5), first consider the following lemma.

Lemma 3.3 Let $G = (V, E, w)$ denote a connected graph of diameter 2^h. For the HDS $\bar{H} = (H_0, H_1, \ldots, H_h)$ obtained by algorithm 3.1 and the arbitrary vertex $v \in V$, if v is not α-padded in $\bar{H}$, there exists some $i \le h - 1$ such that:

$$d_G(v, v.\text{rep}_i) > U \cdot 2^{i-1} - \alpha \cdot 2^i \wedge B_G\left(v, \alpha \cdot 2^{i+1}\right) \subseteq C_v^{(i+1)}$$

where $C_v^{(i+1)}$ is the only $(i + 1)$-level cluster containing v, $v \cdot \text{rep}_i$ denotes the representative of vertex v at level i in algorithm 3.1 ($\forall i < h$), $\alpha \le 1/8$, and $\mathcal{U}$ is the random variable computed in line 2 of algorithm 3.1.

Proof. By contradiction, assume that if v is not an α-padded vertex in $\bar{H}$, then the following proposition holds:

$$\forall i \le h - 1: d_G(u_i, v) \le \mathcal{U} \cdot 2^{i-1} - \alpha \cdot 2^i \vee B_G\left(v, \alpha \cdot 2^{i+1}\right) \nsubseteq C_v^{(i+1)} \quad (3.6)$$

where $u_i = v \cdot \text{rep}_i$. Since v is not α-padded in $\bar{H}$, for some $k < h$, there exists cluster $C \in H_k$ such that $v \in C$ and $B_G(v, \alpha \cdot 2^k) \nsubseteq C$. In addition, let H_k be the highest level partition that contains such a cluster, i.e.:

$$\forall i = k+1, k+2, \ldots, h: \forall C_i \in H_i: B_G\left(v, \alpha \cdot 2^i\right) \subseteq C_i \vee B_G\left(v, \alpha \cdot 2^i\right) \cap C_i = \varnothing \quad (3.7)$$

Since $B_G(v, \alpha \cdot 2^k) \nsubseteq C$, there is a vertex $v' \in B_G(v, \alpha \cdot 2^k)$ such that $v' \notin C$. Let $u_k' \in V$ denote the representative of vertex v' at level k. To the extent that vertex v' does not belong to C, u_k and u_k' are two distinct vertices of G. Assuming cluster $C' = C_v^{(k+1)} \in H_{k+1}$ as the only cluster of level $(k + 1)$ that $C \subseteq C'$ (and subsequently $v \in C'$), we will show that:

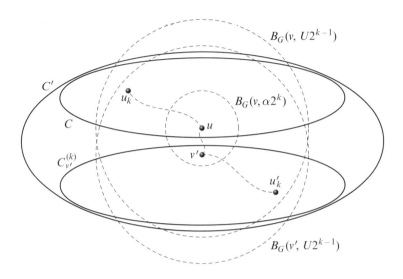

$$u'_k \in C' \cap B_G\left(v, \mathcal{U} \cdot 2^{k-1}\right) \wedge u_k \in C' \cap B_G\left(v', \mathcal{U} \cdot 2^{k-1}\right) \tag{3.8}$$

which contradicts the assumption that $u_k = v \cdot \mathrm{rep}_k$ and $u'_k = v' \cdot \mathrm{rep}_k$ (if u_k appears earlier than u'_k in permutation π, concerning relation (3.8), $u_k = v' \cdot \mathrm{rep}_k$; otherwise, $u'_k = v \cdot \mathrm{rep}_k$). Henceforth, to complete the proof, we only need to prove relation (3.8).

By setting $i = k+1$ in proposition (3.7), we obtain the following relation:

$$B_G\left(v, \alpha \cdot 2^{k+1}\right) \subseteq C' \tag{3.9}$$

As a result, since $v' \in B_G(v, \alpha \cdot 2^k)$, vertex v' is also a member of set C'. Henceforth, concerning line 9 of algorithm 3.1, $u'_k \in C' \cap B_G\left(v', \mathcal{U} \cdot 2^{k-1}\right)$, and $u_k \in C' \cap B_G\left(v, \mathcal{U} \cdot 2^{k-1}\right)$. proposition (3.6) and relation (3.9) imply the following inequality:

$$d_G\left(u_k, v\right) \le \mathcal{U} \cdot 2^{k-1} - \alpha \cdot 2^k$$

Since v' is in $B_G(v, \alpha \cdot 2^k)$, it can be inferred that:

$$d_G\left(v', u_k\right) \le d_G\left(v', v\right) + d_G\left(v, u_k\right) \le \alpha \cdot 2^k + \left(\mathcal{U} \cdot 2^{k-1} - \alpha \cdot 2^k\right) \le \mathcal{U} \cdot 2^{k-1}$$

In addition, since v and v' are not in the same k-level cluster and $v \in B_G(v', \alpha \cdot 2^k)$, $B_G(v', \alpha \cdot 2^k)$ is not a subset of the k-level cluster containing v'. This implies that v' is also not an α-padded vertex in $\bar{H}$. As a result, proposition (3.6) is also true for v' and u'_k, i.e.:

$$d_G\left(u'_k, v'\right) \le \mathcal{U} \cdot 2^{k-1} - \alpha \cdot 2^k$$

As long as vertex v' belongs to $B_G(v, \alpha \cdot 2^k)$, v also belongs to $B_G(v', \alpha \cdot 2^k)$. Subsequently,

$$d_G(v, u'_k) \leq d_G(v, v') + d_G(v', u'_k) \leq \alpha \cdot 2^k + (\mathcal{U} \cdot 2^{k-1} - \alpha \cdot 2^k) \leq \mathcal{U} \cdot 2^{k-1} \quad \blacksquare$$

Now, we start proving inequality (3.5) for every $v \in V$. Concerning the above lemma, we infer that:

$$\Pr[v \text{ is not } \alpha\text{-padded in } \bar{H}] \leq \Pr[\exists i \leq h - 1: d_G(v \cdot rep_i, v) > r_i]$$

Or equivalently,

$$\Pr[v \text{ is not } \alpha\text{-padded in } \bar{H}]$$
$$\leq \Pr[\exists i \leq h - 1: d_G(u, v) > \mathcal{U} \cdot 2^{i-1} - \alpha \cdot 2^i \wedge u = v \cdot rep_i] \qquad (3.10)$$

Note that in inequality (3.10), for every $i < h$, we deduce that:

$$d_G(v, v \cdot rep_i) > \mathcal{U} \cdot 2^{i-1} - \alpha \cdot 2^i \xrightarrow{u \geq \frac{1}{2}} d_G(v, v \cdot rep_i) > 2^{i-2} - \alpha \cdot 2^i$$
$$\xrightarrow{\alpha \leq \frac{1}{8}} d_G(v, v \cdot rep_i) > 2^{i-2} - 2^{i-3} > 0$$

which implies that $v \neq v \cdot rep_i$ for every $i < h$.

Now, consider the following partition of $V - \{v\}$

$$V - \{v\} = \bigcup_{j=1}^{h} B_G(v, 2^j) - B_G(v, 2^{j-1})$$

Using this partition, we rewrite inequality (3.10) in the following form:

$$\Pr[v \text{ is not } \alpha\text{-padded in } \bar{H}] \leq \Pr[\bigvee_{j=1}^{h}(u \in (B_G(v, 2^j) - B_G(v, 2^{j-1})))$$
$$\wedge (\exists i \leq h - 1: d_G(u, v)$$
$$> \mathcal{U} \cdot 2^{i-1} - \alpha \cdot 2^i \wedge u = v \cdot rep_i)]$$

According to algorithm 3.1, since $d_G(v, v \cdot rep) \leq \mathcal{U} \cdot 2^{i-1} \leq 2^{i-1}$ (for every $i \leq h - 1$), we obtain the following inequalities ($r_i = \mathcal{U} \cdot 2^{i-1} - \alpha \cdot 2^i$):

$$\Pr[v \text{ is not } \alpha\text{-padded in } \bar{H}]$$
$$\leq \Pr[\bigvee_{j=2}^{h-1}(u \in (B_G(v, 2^{j-1}) - B_G(v, 2^{j-2}))) \wedge (\exists i \leq h - 1: d_G(u, v) > r_i \wedge u = v \cdot rep_i))]$$
$$\leq \sum_{j=2}^{h-1} \Pr[u \in (B_G(v, 2^{j-1}) - B_G(v, 2^{j-2})) \wedge (\exists i \leq h - 1: d_G(u, v) > r_i \wedge u = v \cdot rep_i)]$$
$$\leq \sum_{j=2}^{h-1} \Pr[\exists i \leq h - 1: d_G(u, v) > r_i \wedge u = v \cdot rep_i \mid u \in (B_G(v, 2^{j-1}) - B_G(v, 2^{j-2}))]$$

In the above inequalities, we will find a relation between integers i and j. As $u = v \cdot rep_i$, u is a member of $B_G(v, \mathcal{U} \cdot 2^{i-1})$. This implies that:

$$\mathcal{U} \cdot 2^{i-1} - \alpha \cdot 2^i < d_G(u, v) \leq \mathcal{U} \cdot 2^{i-1} \xrightarrow{\frac{1}{2} \leq \mathcal{U} < 1}$$
$$(1 - 4\alpha)2^{i-2} < d_G(u, v) < 2^{i-1}$$

Moreover, since u is also in $B_G(v, 2^{j-1}) - B_G(v, 2^{j-2})$, we obtain the following relation between i and j:

$$2^{j-2} < d_G(u,v) \le 2^{j-1} \to \begin{cases} (1-4\alpha)2^{i-2} < 2^{j-1} & \text{if } j \ge j \\ 2^{j-2} < 2^{i-1} & \text{otherwise} \end{cases}$$

$$\to \begin{cases} i < j + \log_2 \dfrac{2}{1-4\alpha} & \text{if } j \ge j \\ i > j-1 & \text{otherwise} \end{cases}$$

Regarding the assumption that $\alpha \le 1/8$, we obtain that:

$$j \le i \le j+2$$

Subsequently, we obtain the following upper bound for $\mathbf{Pr}[v \text{ is not } \alpha\text{-padded in } \bar{H}]$:

$\mathbf{Pr}[v \text{ is not } \alpha\text{-padded in } \bar{H}]$

$$\le \sum_{j=2}^{h-1} \mathbf{Pr}\left[\bigvee_{\substack{i < h \\ j \le i \le j+2}} (d_G(u,v) > U \cdot 2^{i-1} - \alpha \cdot 2^i \wedge u = v \cdot \mathrm{rep}_i) \Big| u \in (B_G(v, 2^{j-1}) - B_G(v, 2^{j-2})) \right]$$

$$\le \sum_{j=2}^{h-1} \sum_{\substack{i < h \\ j \le i \le j+2}} \mathbf{Pr}[d_G(u,v) > U \cdot 2^{i-1} - \alpha \cdot 2^i \wedge u = v \cdot \mathrm{rep}_i \mid u \in B_G(v, 2^{j-1}) - B_G(v, 2^{j-2})]$$

Assuming C_v^{i+1} is the only cluster in H_{i+1} that contains v, the value of the conditional probability:

$$\mathbf{Pr}[d_G(u,v) > U \cdot 2^{i-1} - \alpha \cdot 2^i \wedge u = v \cdot \mathrm{rep}_i \mid u \in (B_G(v, 2^{j-1}) - B_G(v, 2^{j-2}))],$$

is equal to the probability of the following event considering the assumption that $u \in B_G(v, 2^{j-1}) - B_G(v, 2^{j-2})$ (see algorithm 3.1):

Vertex u is the first vertex (concerning permutation π) which belongs to set $B_G(v, U \cdot 2^{i-1}) \cap C_v^{i+1}$ and also satisfies the following equation.

$$U \cdot 2^{i-1} - \alpha \cdot 2^i < d_G(u,v) \le U \cdot 2^{i-1} \qquad (3.11)$$

Note that regarding Lemma 3.3, $B_G(v, 2^{i-2}) \subseteq C_v^{i+1}$ which implies that set $B_G(v, U \cdot 2^{i-1}) \cap C_v^{i+1}$ is equal to $B_G(v, U \cdot 2^{i-1})$.

Inequality (3.11) implies that:

$$\frac{d_G(u,v)}{2^{i-1}} \le U < \frac{d_G(u,v)}{2^{i-1}} + 2\alpha$$

Additionally, as U is uniformly distributed over interval $[1/2, 1)$, we deduce that:

$$\mathbf{Pr}[U = x] \le 2dx \quad \forall x \in \mathbb{R}$$

Moreover, the probability that a vertex appears earlier than n other vertices in the uniformly random permutation π is equal to $1/(n+1)$. Consequently,

$$\mathbf{Pr}\left[d_G(u, v) > U \cdot 2^{i-1} - \alpha \cdot 2^i \wedge u = v \cdot \mathrm{rep}_i \,\middle|\, u \in (B_G(v, 2^{j-1}) - B_G(v, 2^{j-2}))\right]$$

$$= \int_{\frac{d_G(u,v)}{2^{i-1}}}^{\frac{d_G(u,v)}{2^{i-1}}+2\alpha} \frac{1}{\left|B_G\left(v, x \cdot 2^{i-1}\right) \cap C_v^{i+1}\right|} \cdot \mathbf{Pr}\left[\mathcal{U} = x\right]$$

$$= \int_{\frac{d_G(u,v)}{2^{i-1}}}^{\frac{d_G(u,v)}{2^{i-1}}+2\alpha} \frac{1}{\left|B_G\left(v, x \cdot 2^{i-1}\right)\right|} \cdot \mathbf{Pr}\left[\mathcal{U} = x\right]$$

$$\leq \int_{\frac{d_G(u,v)}{2^{i-1}}}^{\frac{d_G(u,v)}{2^{i-1}}+2\alpha} \frac{1}{\left|B_G\left(v, x \cdot 2^{i-1}\right)\right|} \cdot 2\, dx$$

$$\leq \frac{1}{\left|B_G\left(v, \dfrac{d_G(u, v)}{2^{i-1}} \cdot 2^{i-1}\right)\right|} \int_{\frac{d_G(u,v)}{2^{i-1}}}^{\frac{d_G(u,v)}{2^{i-1}}+2\alpha} 2\, dx$$

$$= \frac{4\alpha}{\left|B_G(v, d_G(u, v))\right|}$$

$$\leq \frac{4\alpha}{\left|B_G(v, 2^{j-2})\right|}$$

The last inequality leads to the following inequalities:

$$\mathbf{Pr}\left[v \text{ is not } \alpha\text{-padded in } \bar{H}\right] \leq \sum_{j=2}^{h-1} \sum_{\substack{i < h \\ j \leq i \leq j+2}} \frac{4\alpha}{\left|B_G(v, 2^{j-2})\right|}$$

$$\leq \sum_{j=2}^{h-1} \frac{12\alpha}{\left|B_G(v, 2^{j-2})\right|}$$

$$\leq 12\alpha \sum_{n=1}^{|V|} \frac{1}{n}$$

$$\leq 12\alpha \int_1^{|V|+1} \frac{dx}{x}$$

$$\leq (12\ln 2) \cdot \alpha \log(|V| + 1)$$

Or equivalently, we obtain inequality (3.5) for every $v \in V$.

3.1.3   The Routing Scheme Construction

Now, we are ready to address how to construct a versatile routing scheme for the oblivious routing problem specified in section 3.1.1. At first, we present some preliminary definitions.

Assume that p_1 denotes a walk from v_1 to v_2 and p_2 denotes a walk from v_2 to v_3 in graph $G = (V, E, w)$. The merge operator $\oplus$ is defined on p_1 and p_2 and results in another walk represented by $p_1 \oplus p_2$ in graph G. Walk $p_1 \oplus p_2$ is a walk in G and is obtained by moving from v_1 to v_2 using p_1 and continuing the way to v_3 on walk p_2.

Definition 3.2   Consider the HDS $\bar{H}$ of graph $G = (V, E, w)$ and its corresponding HDT $T = (V_T, E_T)$. The *representative of tree vertex* (C, i) is defined as an arbitrary graph vertex belonging to cluster C $(\forall (C, i) \in V_T)$.

According to Definition 3.2, every tree vertex of an HDT has some representative in the associated graph. More specifically, for each leaf of an HDT in the form $(C, 0)$, its representative is defined as the only member of basic cluster C (note that since graph G satisfies condition (3.1), each basic cluster contains only one vertex.)

Definition 3.3 Consider the HDT T of graph $G = (V, E, w)$ and its corresponding HDS $\bar{H} = (H_1, H_2, \ldots, H_k)$. For every path p of tree T, the *projection of path p* on graph G is defined as the following path in graph G:

$$\text{projection}(p) = \oplus_{\{u,v\}\in p} \text{SP}_G(u', v')$$

such that $\text{SP}_G(u', v')$ represents the shortest path between vertices u' and v' in graph G; and also, $u', v' \in V$ denote the representatives of tree vertices u and v, respectively.

Algorithm 3.2 computes the projection of any path of an HDT on its corresponding graph (note that in this algorithm, for every vertex v of HDT T, *v·rep* specifies its representative vertex in the associated graph).

Fractional Scheme The manner in which we construct a fractional versatile routing scheme for the aforementioned oblivious routing problem is as follows: at first, we use algorithm 3.1 to generate $c \cdot \log|V|$ HDSs for the given connected graph $G = (V, E, w)$ (shortly, we will discuss the value of c). Then, for every generated HDS, we compute its corresponding HDT. For every pair of vertices $u, v \in V$, we mark $\log|V|$ HDTs (among $c \cdot \log|V|$ generated HDTs) in which u and v are both α-padded (for some $\alpha \leq \alpha_0 = \Omega\left(\dfrac{1}{\log|V|}\right)$).

Algorithm 3.2   Projection

Input: Path p which is between two leaves of HDT T and the graph in which tree T is defined.

Output: The projection of tree path p on graph G

1    $p_G \leftarrow \varnothing$;
2   **foreach** tree edge $\{u, v\} \in p$ **do**
3        $p' \leftarrow$ the shortest path in G from vertex $u.rep$ to vertex $v.rep$;
4        $p_G \leftarrow p_G \oplus p'$;
5   **end**
6   **return** p_G;

Assuming that sequence $T_1, T_2, \ldots, T_{\log|V|}$ represents the marked HDTs, the suggested fractional flow between any pair of vertices u and v is obtained by the following equation:

$$\mathbb{S}(s,t) = \left\{ \left(q_i, \frac{1}{\log|V|} \right) : q_i = \text{projection}\left(p_{s,t}^{(i)} \right), i = 1, 2, \ldots, \log|V| \right\}$$

where $p_{s,t}^{(i)}$ denotes the path between leaves $(\{s\}, 0)$ and $(\{t\}, 0)$ in HDT T_i (for every $i = 1, 2, \ldots, \log|V|$). Note that in the above equation, we divide the flow equally into $\log|V|$ paths between s and t.

Algorithm 3.3 describes how to generate a fractional routing scheme in detail. Note that since the input graph of the algorithm satisfies condition (3.1), every leaf of an HDT tree of the graph is in the form $(C, 0)$ such that $|C| = 1$.

Additionally, in algorithm 3.3, it has been implicitly assumed that for every two graph vertices, there are $\log|V|$ HDSs in which both of the vertices are α-padded. We will address this claim thoroughly in the remainder of this section.

Previously, we introduced a randomized algorithm (originally presented by Fakcharoenphol et al., 2003) that receives a weighted connected graph as input and generates a random HDS in which any graph vertex is α-padded with high probability ($\alpha \le 1/8$). Now, consider the following theorem regarding this algorithm:

Theorem 3.1 Let $G = (V, E, w)$ denote a connected graph. Additionally, assume that running algorithm 3.1 for $n = c \cdot \log|V|$ times yields the following HDSs as outputs: $\bar{H}^{(1)}, \bar{H}^{(2)}, \ldots, \bar{H}^{(n)}$ ($c \ge 2$). There is real number $\alpha_0 = \Omega\left(\frac{1}{\log|V|} \right)$ such that: for every pair of vertices $u, v \in V$, with the following probability, there exists at least $\log|V|$ HDSs (among the n HDSs) in which u and v are both α-padded for every $\alpha \le \min\{\alpha_0, 1/8\}$.

Proof. Let U_i and V_i respectively denote the following events ($\forall i = 1, 2, \ldots, n$):

U_i: "Vertex u is α-padded in $\bar{H}^{(i)}$"

V_i: "Vertex v is α-padded in $\bar{H}^{(i)}$"

Concerning inequality (3.5), there exists some $\alpha_0 = \Omega\left(\frac{1}{\log|V|} \right)$ such that for every $\alpha \le \min\{\alpha_0, 1/8\}$, the following inequalities hold:

$$\begin{cases} \mathbf{Pr}\left[\overline{U_i} \right] \le p \\ \mathbf{Pr}\left[\overline{V_i} \right] \le p \end{cases} \quad \forall i = 1, 2, \ldots, n$$

where p is a real number in interval $(0, 1/2)$. This implies that:

Algorithm 3.3   Top-down routing scheme generator (fractional)

Input: Connected graph $G = (V, E, w)$ of diameter 2^h

Output: Versatile routing scheme $\mathbb{S}$

```
1   for  i←1  to   c log|V|  do
2   |     H̄^(i) ← RandomizedHDSGeneraor (G);
3   |     T^(i) ← the HRT corresponding to H̄^(i);
4   end
5   n ← log|V|;
6   foreach  s ∈ V  do
7   |    foreach  t ∈ V − {s}  do
8   |    |    for  i ← 1  to   c. log|V|  do
9   |    |    |    if  n = 0  then
10  |    |    |    |   break;
11  |    |    |    end
12  |    |    |    if  s and t are α-padded in H̄^(i)  then
13  |    |    |    |    mark tree T^(i) as a "padding tree";
14  |    |    |    |    n ← n − 1;
15  |    |    |    end
16  |    |    end
17  |    |    𝕊(s,t) ← ∅;
18  |    |    foreach  padding tree T  do
19  |    |    |    p ← the only path existed between ({s}, 0) and ({t}, 0) in tree T;
20  |    |    |    p_G ← Projection(p, G);
21  |    |    |    𝕊(s,t) ← 𝕊(s,t) ⊕ ( p_G, 1/log|V| );
22  |    |    |    Unmark T;
23  |    |    end
24  |    end
25  end
26  return 𝕊;
```

$$\mathbf{Pr}[U_i \wedge V_i] = \mathbf{Pr}\left[\overline{\overline{U_i} \vee \overline{V_i}}\right] = 1 - \mathbf{Pr}\left[\overline{U_i} \vee \overline{V_i}\right] \geq 1 - \mathbf{Pr}\left[\overline{U_i}\right] - \mathbf{Pr}\left[\overline{V_i}\right] \geq 1 - 2p$$

Henceforth, assuming that $\mathbf{Pr}[U_i \wedge V_i]$, we obtain the following inequality:

$$q \geq 1 - 2p \tag{3.12}$$

Now, considering $(U_1 \wedge V_1)$, $(U_2 \wedge V_2)$, ... , $(U_n \wedge V_n)$ as a sequence of n independent results of a Bernoulli trial and X as a random variable which counts the number of *true* results, X has binomial distribution $B^3(n, q)$. As a result,

$\mathbf{Pr}[\text{there exist at least } \log|V| \text{ HDSs in which } u \text{ and } v \text{ are } \alpha\text{-padded}]$
$\geq 1 - F_X(\log|V| - 1)$

such that F_X is the cumulative distribution function of variable X. Here, we use inequality (3.12) and the Hoeffding's inequality to find an upper bound for the value of $F_X(\log|V| - 1)$:

$$F_X(\log|V| - 1) \leq F_X(\log|V|)$$
$$\leq e^{-2\frac{(qc-1)^2}{c}\log|V|}$$
$$\leq e^{-2\frac{(c-2pc-1)^2}{c}\log|V|}$$

Consequently, by letting $p = 1/4$, we obtain the following inequality:

$\mathbf{Pr}[\text{there exist at least } \log|V| \text{ HDSs in which } u \text{ and } v \text{ are } \alpha\text{-padded}]$
$$\geq 1 - e^{-\frac{(c-2)^2}{2c}\log|V|}$$

Or, equivalently ($n = c\log|V|$),

$\mathbf{Pr}[\text{there exist at least } \log|V| \text{ HDSs in which } u \text{ and } v \text{ are } \alpha\text{-padded}]$
$$\geq 1 - e^{-\frac{(n-2\log|V|)^2}{2n}} \tag{3.13} \blacksquare$$

In Theorem 3.1, by increasing the value of c, we can reach the low threshold to a larger probability, i.e., by generating more HDSs ($n = c\log|V|$), the vertices are α-padded with higher certainty. Figure 3.2 shows the scatter plot of the low threshold of padding probability versus the number of HDSs n.

Integral Scheme Now, we make a similar algorithm to generate an integral versatile solution for the aforementioned oblivious routing problem. As can be seen in algorithm 3.4, at first, we make n HDSs using algorithm

3. Binomial distribution with parameters n and q.

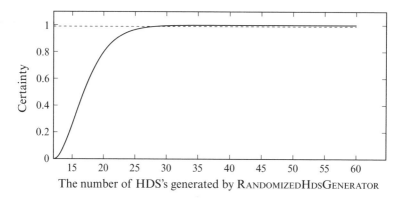

The number of HDS's generated by RANDOMIZEDHDSGENERATOR

Figure 3.2 The plot of the probability low threshold, see inequality (3.13). In this plot, we assume that $\log|V| = 6$, i.e., $n = 6c$. Therefore, if we generate more than 30 HDSs using algorithm 3.1, any pair of vertices will be α-padded in six of them with certainty of more than 99%.

Algorithm 3.4   Top-down routing scheme generator (integral)

Input: Connected graph $G = (V, E, w)$ of diameter 2^h
Output: Versatile routing scheme $\mathbb{S}$

```
1    for  i ← 1  to   c log|V|  do
2    |      H̄^(i) ← RandomizedHDSGeneraor (G);
3    |      T^(i) ← the HRT corresponding to H̄^(i);
4    end
5    foreach  s ∈ V  do
6    |      foreach  t ∈ V − {s}  do
7    |      |      for  i ← 1  to   c. log|V|  do
8    |      |      |      if  s and t are α-padded in H̄^(i)  then
9    |      |      |      |      T ← T^(i);
10   |      |      |      |      break;
11   |      |      |      end
12   |      |      end
13   |      |      p ← the only path existed between ({s}, 0) and ({t}, 0) in
     |      |            tree T;
14   |      |      S(s, t) ← Projection (p, G);
15   |      end
16   end
17   return  S;
```

3.1; then, we compute the associated n HDTs of the generated HDSs. For every pair of vertices s and t, we will find an HDT in which the vertices are both α-padded (similar to the fractional version, we can make the value of n large enough to make sure that there is such an α-padding HDT). Finally, the suggested path between vertices s and t in the input graph is the projection of the only path between leaves ($\{s\}$, 0) and ($\{t\}$, 0) in the α-padding HDT.

Figure 3.3 represents a path suggested by algorithm 3.4 through a weighted grid graph $G_{5\times5} = (V, E, w)$ such that $w(e) = 2$ for every $e \in E$. As it is shown, an HDT of the graph has been depicted schematically. In this figure, each tree vertex is labeled by the set of vertices belonging to its corresponding cluster. The underlined numbers specify the representatives of each tree vertex in the grid graph. Additionally, the way of projecting a tree path on the graph has been specified.

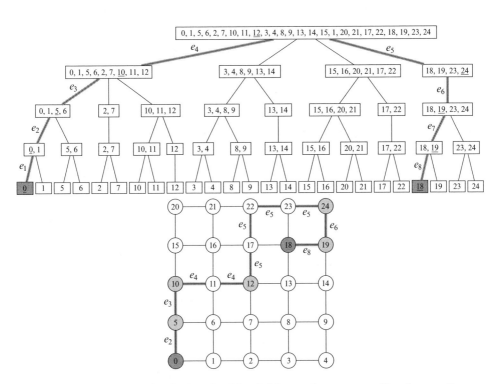

Figure 3.3 An example of using algorithm 3.4 for routing a commodity of source 0 and target 18 through the shown grid graph ($G_{5\times5}$). Note that the weight of each edge in the grid is assumed to be 2.

3.1.4 Routing Cost Analysis

As mentioned in chapter 1, in order to analyze the cost efficiency of a versatile routing scheme, we use the competitive ratio, which is defined as the maximum cost incurred by the versatile routing scheme divided by the routing cost of the optimal solution. Here, we first find a lower bound for the cost of the optimal solution (whether it is fractional or integral). Then, we compute a high threshold for the routing cost of the solution suggested by the schemes described previously. Finally, we find the competitive ratios of both the fractional and integral routing schemes.

Here, we find a lower bound for the cost of the optimal solution of the aforementioned oblivious routing problem. At first, we introduce the concepts of α-padding cover and cutting a commodity in a weighted graph.

Definition 3.4 Let $G = (V, E, w)$ denote a connected graph. Assume that $\mathcal{H}$ represents a set of $\log|V|$ HDSs of graph G and $\mathcal{E}$ denotes a routing cost environment in graph G. Set $\mathcal{H}$ is called an *α-padding cover in environment $\mathcal{E}$* if the following condition holds:

$$\forall i = 1, 2, \ldots, k : \exists \bar{H} \in \mathcal{H} \text{ such that vertices } \Pi_1(K)$$
$$\text{and } \Pi_2(K) \text{ are α-padded in HSD } \bar{H}$$

Furthermore, HDS $\bar{H}$ cuts commodity K in the ith level if $i \le h - 1$:

$$\exists C \in \Pi_i(\bar{H}): \Pi_1(K) \in C \wedge \Pi_2(K) \notin C$$

Additionally, a *closed cut set* of a commodity is defined as follows.

Definition 3.5 Consider a general routing problem of weighted graph $G = (V, E, w)$ and routing cost environment $\mathcal{E}$. Set $X \subseteq E$ is called a *closed cut set* of the ith commodity in environment $\mathcal{E}$, if:

i. there exists an integral solution like $\bar{P}$ such that:

$$\sum_{e \in X} f_{\bar{P}}^{(i)}(e) = \Pi_3(K_i)$$

ii. and there is no path between vertices $\Pi_1(K_i)$ and $\Pi_2(K_i)$ in graph $G^X = (V, E - X)$

where $(K_1, K_2, \ldots, K_k)$ denotes the sequence of commodities in environment $\mathcal{E}$.

Now, consider the following theorem, which presents a lower bound for the solution cost of the aforementioned routing problem.

Theorem 3.2 Consider the oblivious routing problem of graph $G = (V, E, w)$ of diameter 2^h and set of possible routing cost environments $\mathbb{E}$, as defined in equation (3.3). If function S denotes an arbitrary solution of the problem (either integral or fractional) and set $\mathcal{H}_{\mathcal{E}}$ represents an α-padding cover in environment $\mathcal{E}$ (for every $\mathcal{E} \in \mathbb{E}$), the following inequality holds:

$$C_{\mathcal{E}}(S(\mathcal{E})) \geq \Theta\left(\frac{\alpha}{4|\mathcal{H}_{\mathcal{E}}|}\sum_{i=0}^{h-1}\sum_{\bar{H}\in\mathcal{H}_{\mathcal{E}}} 2^i \cdot \mathrm{rrc}\big(b_{1,i}(\bar{H}), b_{2,i}(\bar{H}), \ldots, b_{k,i}(\bar{H})\big)\right) \quad \forall \mathcal{E} \in \mathbb{E}$$

where $\alpha = \min(1/4, \alpha_0)$ for some $\alpha_0 = \Omega(\log|V|)$ and $b_{n,i}(\bar{H})$ is defined in the following way:

$$b_{n,i}(\bar{H}) = \begin{cases} \Pi_3(K_n) & \begin{array}{l}\text{if } \Pi_1(K_n) \text{ and } \Pi_2(K_n) \text{ are } \alpha\text{-padded in} \\ \bar{H} \text{ and } \bar{H} \text{ cuts } K_n \text{ in the } i\text{th level}\end{array} \\ 0 & \text{otherwise} \end{cases}$$

and $(K_1, K_2, \ldots, K_k)$ denotes the sequence of commodities in environment $\mathcal{E}$.
Proof. Since S is the solution function of the oblivious routing problem, the network routing cost is computed by the following equation:

$$C_{\mathcal{E}}(S(\mathcal{E})) = \mathrm{nrc}_\pi\left(\begin{array}{c}\mathrm{cost}_{\pi_1}(f_{11}, f_{12}, \ldots, f_{1k}), \mathrm{cost}_{\pi_2}(f_{21}, f_{22}, \ldots, f_{2k}), \ldots, \\ \mathrm{cost}_{\pi_{|E|}}(f_{|E|1}, f_{|E|2}, \ldots, f_{|E|k})\end{array}\right) \quad \forall \mathcal{E} \in \mathbb{E}$$

where $f_{i,j}$ denotes the traffic flow value of some specific commodity in the ith edge (concerning permutation π):

$$f_{i,j} = f_{S(\mathcal{E})}^{(j)}(\pi_i) \quad \forall i = 1, 2, \ldots, |E|, \ \forall j = 1, 2, \ldots, k$$

Concerning equation (3.3), we can simplify the cost function in the following way:

$$C_{\mathcal{E}}(S(\mathcal{E})) = \sum_{e\in E} w(e) \cdot \mathrm{rrc}\big(f_{S(\mathcal{E})}^{(1)}(e), f_{S(\mathcal{E})}^{(2)}(e), \ldots, f_{S(\mathcal{E})}^{(k)}(e)\big) \quad \forall \mathcal{E} \in \mathbb{E}$$

Additionally, since $\mathcal{H}_{\mathcal{E}}$ denotes some α-padding cover in environment $\mathcal{E} \in \mathbb{E}$, we infer that:

$$C_{\mathcal{E}}(S(\mathcal{E})) = \frac{1}{|\mathcal{H}_{\mathcal{E}}|}\sum_{\bar{H}\in\mathcal{H}_{\mathcal{E}}}\sum_{e\in E} w(e) \cdot \mathrm{rrc}\big(f_{S(\mathcal{E})}^{(1)}(e), f_{S(\mathcal{E})}^{(2)}(e), \ldots, f_{S(\mathcal{E})}^{(k)}(e)\big)$$
$$\geq \frac{1}{|\mathcal{H}_{\mathcal{E}}|}\sum_{\bar{H}\in\mathcal{H}_{\mathcal{E}}}\sum_{e\in E} w(e) \cdot \mathrm{rrc}\big(g_1(e, \bar{H}), g_2(e, \bar{H}), \ldots, g_k(e, \bar{H})\big)$$

$$\forall \mathcal{E} \in \mathbb{E}$$

where:

$$g_n(e, \bar{H}) = \begin{cases} f_{S(\mathcal{E})}^{(n)}(e) & \text{if } \Pi_1(K_n) \text{ and } \Pi_2(K_n) \text{ are } \alpha\text{-padded in } \bar{H} \\ 0 & \text{otherwise} \end{cases} \tag{3.14}$$
$$\forall n = 1, 2, \ldots, k$$

and $(K_1, K_2, \ldots, K_k)$ represents the sequence of commodities in environment $\mathcal{E}$.

Now, we consider the following subset of edge set E for every $\bar{H} \in \mathcal{H}$:

$$\bigcup_{i=0}^{h-1}\bigcup_{C\in\Pi_i(\bar{H})} E_{i,C} \subseteq E \tag{3.15}$$

such that $Ei, c = \varnothing$ if C does not contain a commodity terminal (source and target) or if C contains both terminals of a single commodity. On the other hand, if there exists some commodity terminal in C and $\bar{H}$ cuts it in the ith level, $E_{i,C}$ is obtained from the following equation:

$$E_{i,C} = \{\{u, v\} \in E : \{u, v\} \subseteq C \wedge \max\{m_u, m_v\} \in (\alpha \cdot 2^{i-1}, \alpha \cdot 2^i]\}$$

where m_u denotes the minimum graph distance between u and any commodity terminal which belongs to C (similar definition for m_v). See figure 3.4 for an illustration.

According to relation (3.15), we obtain the following lower bound for the solution routing cost:

$$C_{\mathcal{E}}(S(\mathcal{E}))$$
$$\geq \frac{1}{|\mathcal{H}_{\mathcal{E}}|} \sum_{\bar{H} \in \mathcal{H}_{\mathcal{E}}} \sum_{i=0}^{h-1} \sum_{C \in \Pi_i(\bar{H})} \sum_{e \in E_{i,C}} w(e) \cdot \mathrm{rrc}\big(g_1(e, \bar{H}), g_2(e, \bar{H}), \ldots, g_k(e, \bar{H})\big)$$
$$\forall \mathcal{E} \in \mathbb{E}$$

Now, we construct graph $G' = (V', E', w')$ using graph $G = (V, E, w)$ in the following algorithm:

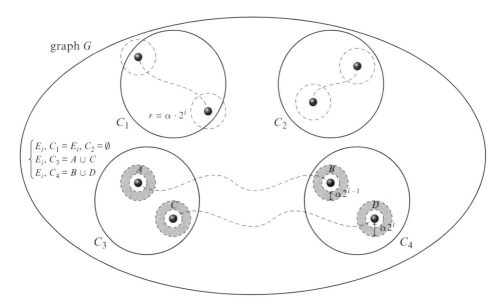

Figure 3.4 A schematic view of graph G and an HDS at level i. As can be seen, cluster C_1 does not contain any α-padded commodity terminal ($E_{i,C_1} = \varnothing$). Additionally, the pair of α-padded vertices in C_2 are terminals of one commodity (subsequently, $E_{i,C_2} = \varnothing$). However, there are α-padded terminals in clusters C_3 and C_4 such that the shown HDS has cut their corresponding commodities in the ith level.

At first, assign sets $\emptyset$ and V to E' and V', respectively. Then, for every edge $e \in E$, add the following $(\lfloor w(e) \rfloor - 1)$ vertices to set V': $x_{1e}, x_{2e}, \ldots, x_{(\lfloor w(e) \rfloor - 1)e}$. Additionally, add the following $\lfloor w(e) \rfloor$ edges to set E':

$$\{u, x_{1e}\}, \{x_{1e}, x_{2e}\}, \ldots, \{x_{(\lfloor w(e) \rfloor - 1)e}, v\} \tag{3.16}$$

where $e = \{u, v\}$. For every $e' \in E'$, assign $w'(e') = 1$.

Note that in this algorithm, the maximum error of distance function $d_{G'}$ for any pair of adjacent vertices in V is 1 (in comparison with function d_G). We can scale up the weight function w so that this error becomes zero, i.e.:

$$d_G(u, v) = d_{G'}(u, v) \quad \forall u, v \in V$$

Every solution of the oblivious routing problem in graph G can be mapped to G' by mapping the flow of any edge $e = \{u, v\}$ of graph G to its corresponding sequence of edges, as noted in equation (3.16). In the exercises, you will be asked to prove that the routing cost of the mapped solution in G' is lower than its associated cost in G. As a result, we obtain that:

$$C_{\mathcal{E}}(S(\mathcal{E}))$$

$$\geq \frac{1}{|\mathcal{H}_{\mathcal{E}}|} \sum_{\bar{H} \in \mathcal{H}_{\mathcal{E}}} \sum_{i=0}^{h-1} \sum_{C \in \Pi_i(\bar{H})} \sum_{e \in E'_{i,C}} w(e) \cdot \mathrm{rrc}\left(g_1(e, \bar{H}), g_2(e, \bar{H}), \ldots, g_k(e, \bar{H})\right)$$

$$\forall \mathcal{E} \in \mathbb{E}$$

such that $E'_{i,C} \subseteq E'$ is defined the same as $E_{i,C}$. Consider the following claim regarding set $E'_{i,C}$:

Claim 3.1 For every environment $\mathcal{E} \in \mathbb{E}$, HDS $\bar{H} \in \mathcal{H}_{\mathcal{E}}$, level $i \leq h - 1$, and cluster $C \in \Pi_i(\bar{H})$, the following statement is true:

Assuming K as a commodity of environment $\mathcal{E}$ and set $E'_{i,C} \neq \emptyset$, if K has a terminal in C and is cut by HDS $\bar{H}$ in the ith level, there will exist $\lfloor \alpha \cdot 2^i \rfloor - \lfloor \alpha \cdot 2^{i-1} \rfloor$ closed cut sets of K in graph G' such that they are disjoint subsets of $E'_{i,C}$.

In the exercises, you will be asked to prove Claim 3.1. If $T_{i,C}$ denotes one of the mentioned closed cut sets, we obtain this inequality:

$$C_{\mathcal{E}}(S(\mathcal{E}))$$

$$\geq \frac{1}{|\mathcal{H}_{\mathcal{E}}|} \sum_{\bar{H} \in \mathcal{H}_{\mathcal{E}}} \sum_{i=0}^{h-1} \sum_{C \in \Pi_i(\bar{H})} (\alpha 2^{i-1} - 1) \sum_{e \in T_{i,C}} \mathrm{rrc}\left(g_1(e, \bar{H}), g_2(e, \bar{H}), \ldots, g_k(e, \bar{H})\right)$$

$$\geq \Theta\left(\frac{1}{|\mathcal{H}_{\mathcal{E}}|} \sum_{\bar{H} \in \mathcal{H}_{\mathcal{E}}} \sum_{i=0}^{h-1} \alpha 2^{i-1} \sum_{C \in \Pi_i(\bar{H})} \sum_{e \in T_{i,C}} \mathrm{rrc}\left(g_1(e, \bar{H}), g_2(e, \bar{H}), \ldots, g_k(e, \bar{H})\right)\right)$$

In addition, regarding the assumption that function rrc is sub-additive, we conclude the following inequality:

$$C_{\mathcal{E}}(S(\mathcal{E}))$$

$$\geq \Theta\left(\frac{1}{|\mathcal{H}_{\mathcal{E}}|}\sum_{\bar{H}\in\mathcal{H}_{\mathcal{E}}}\sum_{i=0}^{h-1}\alpha 2^{i-1}\mathrm{rrc}\left(\frac{\sum_{C\in\Pi_i(\bar{H})}\sum_{e\in T_{i,C}}g_1(e,\bar{H}),\sum_{C\in\Pi_i(\bar{H})}}{\sum_{e\in T_{i,C}}g_2(e,\bar{H}),\dots,\sum_{C\in\Pi_i(\bar{H})}\sum_{e\in T_{i,C}}g_k(e,\bar{H})}\right)\right)$$

Since $T_{i,C}$ makes a cut set for every padded commodity terminal inside C, we infer that:

$$b_{n,i}(\bar{H}) = \frac{1}{2}\sum_{C\in\Pi_i(\bar{H})}\sum_{e\in T_{i,C}}g_n(e,\bar{H}) \quad \forall\bar{H}\in\mathcal{H}_{\mathcal{E}}\forall n = 1, 2, \dots, k\forall i \leq h-1 \quad (3.17)$$

Note that the RHS of the above equation has been divided by two because of the fact that each commodity has two terminals. Finally, equation (3.17) implies that:

$$C_{\mathcal{E}}(S(\mathcal{E})) \geq \Theta\left(\frac{\alpha}{4|\mathcal{H}_{\mathcal{E}}|}\right)\sum_{i=0}^{h-1}\sum_{\bar{H}\in\mathcal{H}_{\mathcal{E}}}2^i \cdot \mathrm{rrc}\left(b_{1,i}(\bar{H}), b_{2,i}(\bar{H}), \dots, b_{k,i}(\bar{H})\right) \quad \forall\mathcal{E}\in\mathbb{E}$$

∎

Assume that we focus on the oblivious routing problem of the following set of possible routing cost environments:

$$\mathbb{E}' = \left\{(\mathrm{cost}_e, \Sigma, \bar{K}) \in \mathbb{E}: \mathrm{cost}_e \in F_{\mathrm{commodity\text{-}ind}}\right\} \quad (3.18)$$

where set $\mathbb{E}$ is defined in equation (3.3) and $F_{\mathrm{commodity\text{-}ind}}$ denotes the set of all the functions that hold the following condition:

$$\mathrm{cost}_e(F) = \mathrm{cost}_e(F_1) + \mathrm{cost}_e(F_2) \quad \forall e \in E$$

where F, F_1, and F_2 are sequences of k real positive numbers such that:

$$\Pi_i(F_1) = \begin{cases} \Pi_i(F) & \text{if } \Pi_i(F_2) = 0 \\ 0 & \text{otherwise} \end{cases} \quad \forall i = 1, 2, \dots, k$$

If an edge routing cost function has this property, it is said to be *commodity-independent*, i.e., the routing cost of any commodity does not affect the amount of cost incurred by other commodities. There are many real-world examples that have commodity-independent edge routing cost functions.

By restricting the set of possible routing cost environments to $\mathbb{E}'$, the following equation is deduced from the proof of Theorem 3.2:

$$\mathrm{rrc}\left(f_{S(\mathcal{E})}^{(1)}(e), f_{S(\mathcal{E})}^{(2)}(e), \dots, f_{S(\mathcal{E})}^{(k)}(e)\right)$$
$$= \sum_{\bar{H}\in\mathcal{H}_{\mathcal{E}}}\mathrm{rrc}\left(g_1(e,\bar{H}), g_2(e,\bar{H}), \dots, g_k(e,\bar{H})\right) \quad \forall\mathcal{E}\in\mathbb{E}'$$

such that e is an arbitrary graph edge and $g_n(e,\bar{H})$ is the function defined in equation (3.14). Using a similar deduction to the one we made in the proof of Theorem 3.2, we obtain the following lower bound for the network routing cost of an arbitrary solution to the oblivious routing problem of set $\mathbb{E}'$ of possible routing cost environments:

$$C_{\mathcal{E}}(S(\mathcal{E})) \geq \Theta\left(\frac{\alpha}{4} \sum_{i=0}^{h-1} \sum_{\bar{H} \in \mathcal{H}_{\mathcal{E}}} 2^i \cdot \mathrm{rrc}\left(b_{1,i}(\bar{H}), b_{2,i}(\bar{H}), \dots, b_{k,i}(\bar{H})\right)\right) \quad \forall \mathcal{E} \in \mathbb{E}'$$

(3.19)

Competitive Ratio of the Integral Scheme Finally, in the following theorem, we will find an upper bound for the competitive ratio of the integral routing scheme specified in algorithm 3.4.

Theorem 3.3 If $\mathbb{S}_{\text{integral}}$ denotes the integral versatile routing scheme specified in algorithm 3.4, the competitive ratio of $\mathbb{S}_{\text{integral}}$ in set $\mathbb{E}$ of possible routing cost environments, as defined in equation (3.3), has the following upper bound:

$$\mathrm{CR}\left(\mathbb{S}_{\text{integral}}, \mathbb{E}\right) \leq \Theta\left(\log^2 |V|\right)$$

Additionally, if $\mathbb{E}'$ denotes the set defined in equation (3.18), we will get the following upper bound for the scheme competitive ratio:

$$\mathrm{CR}\left(\mathbb{S}_{\text{integral}}, \mathbb{E}'\right) \leq \Theta\left(\log |V|\right)$$

Proof. At first, we prove the following upper bound for the routing cost of the solution suggested by scheme $\mathbb{S}_{\text{integral}}$ in environment $\mathcal{E} \in \mathbb{E}$ of possible routing cost environments:

$$C_{\mathcal{E}}(S_{int}(\mathcal{E})) \leq \sum_{i=0}^{h-1} \sum_{\bar{H} \in \mathcal{H}_{\mathcal{E}}} 2^i \cdot \mathrm{rrc}\left(b_{1,i}(\bar{H}), b_{2,i}(\bar{H}), \dots, b_{k,i}(\bar{H})\right)$$

(3.20)

such that S_{int} denotes the suggested solution by scheme $\mathbb{S}_{\text{integral}}$ and $b_{n,i}(\bar{H})$ is defined in the following way (for every $n = 1, 2, \dots, k$):

$$b_{n,i}(\bar{H}) = \begin{cases} \Pi_3(K_n) & \text{if } \Pi_1(K_n) \text{ and } \Pi_2(K_n) \text{ are } \alpha\text{-padded in} \\ & \bar{H} \text{ and } \bar{H} \text{ cuts } K_n \text{ in the } i\text{th level} \\ 0 & \text{otherwise} \end{cases}$$

and $(K_1, K_2, \dots, K_K)$ denotes the sequence of commodities in environment $\mathcal{E}$.

Now, we prove inequality (3.20).

$$C_{\mathcal{E}}(S_{int}(\mathcal{E})) = \sum_{e \in E} w(e) \cdot \mathrm{rrc}\left(f_{S(\mathcal{E})}^{(1)}, f_{S(\mathcal{E})}^{(2)}, \dots, f_{S(\mathcal{E})}^{(k)}\right) \quad \forall \mathcal{E} \in \mathbb{E}$$

Let (s, t, val) denote the nth commodity in environment $\mathcal{E} \in \mathbb{E}$. Consider path $p = \mathbb{S}_{\text{integral}}(s, t)$ in graph $G = (V, E, w)$. Regarding algorithm 3.4, path p is the union of a number of paths that are projections of tree edges at different levels of HRT $T^{(m)}$ for some m (see line 3 of the algorithm). Additionally, assume that path $q \subseteq p$ denotes the projection of edge $\{(C, i), (C', i+1)\}$ of tree $T^{(m)}$, $u = (C, i)$, and $u' = (C', i+1)$ for some $i < h$ ($h = \lfloor \log \mathrm{diam}_G \rfloor$). Since u and u' belong to cluster C' and C' is at level $i+1$, we conclude that:

$$d_G(u, u') \leq 2^{i+1}$$

The recent inequality implies that:

$$C_{\mathcal{E}}(S_{\text{int}}(\mathcal{E})) = \sum_{e \in E} w(e) \cdot \text{rrc}\left(f_{S(\mathcal{E})}^{(1)}(e), f_{S(\mathcal{E})}^{(2)}(e), \ldots, f_{S(\mathcal{E})}^{(k)}(e)\right)$$

$$\leq \sum_{i=0}^{h-1} \sum_{\bar{H} \in \mathcal{H}_{\mathcal{E}}} 2^{i+1} \cdot \text{rrc}\left(b_{1,i}(\bar{H}), b_{2,i}(\bar{H}), \ldots, b_{k,i}(\bar{H})\right)$$

The competitive ratio is computed by dividing the upper bound obtained by inequality (3.20) by the lower bound of the routing cost mentioned in Theorem 3.2. Consequently, we find the following high threshold for the competitive ratio of scheme $\mathbb{S}_{\text{integral}}$:

$$CR(\mathbb{S}_{\text{integral}}, \mathbb{E}) \leq \frac{\sum_{i=0}^{h-1} \sum_{\bar{H} \in \mathcal{H}_{\mathcal{E}}} 2^{i+1} \cdot \text{rrc}\left(b_{1,i}(\bar{H}), b_{2,i}(\bar{H}), \ldots, b_{k,i}(\bar{H})\right)}{\Theta\left(\frac{\alpha}{4|\mathcal{H}_{\mathcal{E}}|} \sum_{i=0}^{h-1} \sum_{\bar{H} \in \mathcal{H}_{\mathcal{E}}} 2^{i} \cdot \text{rrc}\left(b_{1,i}(\bar{H}), b_{2,i}(\bar{H}), \ldots, b_{k,i}(\bar{H})\right)\right)}$$

$$\leq \Theta\left(\frac{8|\mathcal{H}_{\mathcal{E}}|}{\alpha}\right)$$

Since $|\mathcal{H}_{\mathcal{E}}| = \Theta(\log|V|)$, by letting $\alpha = \alpha_0 = \Omega\left(\frac{1}{\log|V|}\right)$, we obtain some high threshold asymptotically equal to $\Theta(\log^2|V|)$.

Additionally, in order to find a high threshold for the competitive ratio of the versatile scheme in set $\mathbb{E}'$ of possible routing cost environments, we use inequality (3.19) to obtain the following inequalities:

$$CR(\mathbb{S}_{\text{integral}}, \mathbb{E}') \leq \frac{\sum_{i=0}^{h-1} \sum_{\bar{H} \in \mathcal{H}_{\mathcal{E}}} 2^{i+1} \cdot \text{rrc}\left(b_{1,i}(\bar{H}), b_{2,i}(\bar{H}), \ldots, b_{k,i}(\bar{H})\right)}{\Theta\left(\frac{\alpha}{4} \sum_{i=0}^{h-1} \sum_{\bar{H} \in \mathcal{H}_{\mathcal{E}}} 2^{i} \cdot \text{rrc}\left(b_{1,i}(\bar{H}), b_{2,i}(\bar{H}), \ldots, b_{k,i}(\bar{H})\right)\right)}$$

$$\leq \Theta\left(\frac{8}{\alpha}\right)$$

This implies that there is an upper bound of $\Theta(\log|V|)$ for the competitive ratio if the routing cost environment is restricted to the commodity-independent functions for computing the edge routing costs. ∎

3.2   A Bottom-Up Versatile Routing Scheme

In this section, we provide the bottom-up versatile routing scheme presented by Srinivasagopalan, Busch, and Iyengar in 2012. This routing scheme suggests a versatile solution to the single-source oblivious routing problem.

In chapter 2, three different types of HITs were introduced and addressed in detail. Before starting our discussion on the routing scheme, we summarize some important properties of the HITs. We use the following notation in this chapter:

Consider HIT $T^s = (V_{T^s}, E_{T^s})$ of graph G.

- Assuming level i vertex (v, i) and its parent $(u, i + 1)$ in tree T^s, μ_i denotes *the upper bound of the distance between their corresponding vertices in graph G:*

$$\mu_i = \sup\{d_G(u, v): \{(u, i+1), (v, i)\} \in E_{T^s}\} \quad \forall i \leq h - 1 \qquad (3.21)$$

- For every level i vertex (v, i) of tree T^s, *the upper bound of the graph distance between v and the corresponding vertex of every leaf in subtree $T^s_{(v,i)}$* is represented by δ_i:

$$\delta_i = \sup\{d_G(u, v): u \in \mathcal{L}_{T^s}(v, i), (v, i) \in V_{T^s}\} \quad \forall i \leq h \qquad (3.22)$$

- If (v, i) denotes some vertex in T^s such that $v \neq s$, *the lower bound of the graph distance between source vertex s and the corresponding vertex of every leaf in subtree $T^s_{(v,i)}$* is denoted by ϕ_i:

$$\phi_i = \inf\{d_G(s, u): u \in \mathcal{L}_{T^s}(v, i), (v, i) \in V_{T^s}, v \neq s\} \quad \forall i \leq h \qquad (3.23)$$

Table 3.1 compares the different types of hierarchical independence trees based on parameters μ_i, δ_i, and ϕ_i. Assuming that T^s is an HIT type-0 or type-2, Lemma 2.6 and Lemma 2.8 imply that:

$$B_G(s, 2^i - 1) \subseteq \mathcal{L}_{T^s}(s, i) \quad i \leq h$$

which implies that the graph distance between the source vertex and those that are not in set $\mathcal{L}_{T^s}(s, i)$ is at least 2^i. Additionally, for every vertex $u \notin \mathcal{L}_{T^s}(s, i)$ there is some vertex $v \neq s$ in the ith level that $u \in \mathcal{L}_{T^s}(v, i)$. Hence, we conclude that $\phi_i = 2^i$, and also:

$$\phi_i - 1 = \sup\{d_G(s, v): v \in \mathcal{L}_{T^s}(s, i)\} \quad \forall i \leq h \qquad (3.24)$$

Moreover, figure 3.5 depicts some example which verifies that for an HIT type-1, $\phi_i \leq 2^i$. It is easy to verify the other contents of table 3.1 by considering the HIT definitions in chapter 2, Lemma 2.4, and Lemma 2.6.

3.2.1 Problem Specification

Now, we specify the oblivious routing problem that will have to be solved in a versatile manner. This routing problem is defined in connected

Table 3.1 Comparison of different types of HIT

HIT	$\log_2 \mu_i$	$\log_2 \delta_i$	$\log_2 \phi_i$	HIS type
Type-0	$i+1$	$i+1$	i	Source-oriented
Type-1	$i+1$	$i+1$	$<i$	Basic
Type-2	$i+2$	$i+2$	i	Basic

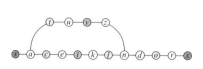
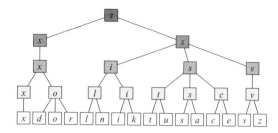

Figure 3.5 Schematic view of an HIS in some connected graph (left), and its corresponding HIT type-1 (right). In this figure, $d \in \mathcal{L}_{T^s}(x, 3)$, but $d_G(s, d) < 2^3$. Moreover, $B_G(s, 2^3 - 1) \not\subseteq \mathcal{L}_{T^s}(s, 3)$.

unweighted graph $G(V, E)$, or, equivalently, weighted graph (V, E, unit). Assuming that $s \in V$ and $D \subseteq V$ denote the *source vertex* and the *set of target vertices*, respectively, the routing process is done in routing cost environment $\mathcal{E} \in \mathbb{E}$ such that $\mathbb{E}$ is defined in the following way:

$$\mathbb{E} = \left\{ (\text{rrc}, \Sigma, \bar{K}) : \text{rrc} \in F_{\text{sub-additive}}, \bar{K} \in \mathcal{K} \right\} \tag{3.25}$$

where $F_{\text{sub-additive}}$ is the set of all the sub-additive functions in the form $\text{rrc} : (\mathbb{Z}_{\geq 0})^{|D|} \mapsto \mathbb{R}_{\geq 0}$ and $\mathcal{K}$ is the set of all the commodity sequences of common source ($s \in V$) and value (1), i.e.:

$$\mathcal{K} = \left\{ (K_1, K_2, \ldots, K_{|D|}) : K_i = (s, t_i, 1), t_i \in D \quad \forall i \in [1, |D|] \right\} \tag{3.26}$$

Note that the network routing cost in this problem is the summation function. For the remainder of this chapter, we represent this problem in the form $P_{s,D}(G, \mathbb{E})$.

3.2.2 Scheme Construction

After specifying the oblivious routing problem, here we use the HIT type-0 as a well-designed hierarchical routing tool to construct a competitive bottom-up routing scheme. Using this scheme, we will find a low-cost solution to the aforementioned problem. At first, we introduce the concept of projecting an HIT path on the corresponding graph of the HIT.

Definition 3.6 Let $T^s = (V_{T^s}, E_{T^s})$ denote an HIT type-0 of height h in graph $G = (V, E)$. For every tree edge of level $i \leq h - 1$ in the form $e = \{(u, i), (v, i + 1)\}$, the *projection of e on graph G* is defined as the shortest path between the vertices u and v in G and denoted by $\text{proj}(e, G)$. In addition, if p represents a path in tree T^s, the projection of p on graph G is defined in the following form:

$$\text{proj}(p, G) = \bigoplus_{e \in p} \text{proj}(e, G) \tag{3.27}$$

For any routing problem in the form $P_{s,D}(G, \mathbb{E})$, the versatile routing scheme suggested by HIT type-0 T^s is defined as the following function:

$$\mathbb{S}_{T^s} \colon (\{s\} \times V) \mapsto (\text{the set of all the paths in } G)$$

such that for every target $t \in D$, the following equation holds:

$$\mathbb{S}_{T^s}(s, t) = \text{proj}(p_{T^s}((t, 0), (s, h)), G)$$

such that $p_{T^s}((t, 0), (s, h))$ denotes the only existing path between the root of T^s and leaf $(t, 0)$. Note that $\mathbb{S}_{T^s}(s, t)$ is not necessarily a simple path in G (as it may cross itself).

Solution Cost Analysis Assume that $D = \{t_1, t_2, \dots, t_{|D|}\}$. Regarding equation (3.25), the cost environment $\mathcal{E} \in \mathbb{E}$ is in the following form:

$$\mathcal{E} = (c(\Sigma), \Sigma, (K_1, K_2, \dots, K_{|D|}))$$

such that for every $i = 1, 2, \dots, |D|$, $K_i = (s, t_i, 1)$. Scheme $\mathbb{S}_{T^s}$ provides a solution of the problem of cost environment $\mathcal{E}$ in the following form:

$$p = (p(t_1), p(t_2), \dots, p(t_{|D|}))$$

where $p(t_j)$ is the projection of the unique tree path between $(t_j, 0)$ and (s, h) on graph G. Moreover, for every $j = 1, 2, \dots, |D|$,

$$p(t_j) = \bigoplus_{i=0}^{h-1} p_i(t_j) \tag{3.28}$$

such that $p_i(t_j)$ is the projection of tree edge $\{\text{parent}_i((t_j, 0)), \text{parent}_{i+1}((t_j, 0))\}$ on G.

Concerning the definition of flow function in chapter 1, we infer that:

$$f_p^{(j)}(e) = \begin{cases} 1 & e \in p(t_j) \\ 0 & \text{otherwise} \end{cases} \quad \forall j = 1, 2, \dots, |D|$$

In addition, assume that $f_p^{(j)}(e, i)$ is defined by the following equation:

$$f_p^{(j)}(e, i) = \begin{cases} 1 & e \in p_i(t_j) \\ 0 & \text{otherwise} \end{cases} \quad \forall i \leq h \ \forall j = 1, 2, \dots, |D| \tag{3.29}$$

Using equation (3.28), we obtain the following inequality for every $j = 1, 2, \dots, |D|$:

$$f_p^{(j)}(e) \leq \sum_{i=0}^{h-1} f_p^{(j)}(e, i) \tag{3.30}$$

Regarding the definition of the edge routing cost function in chapter 1, the routing cost of edge e in solution p is equal to:

$$C_p(e) = c\left(\sum_{j=1}^{|D|} f_p^{(j)}(e)\right)$$

Since c is an increasing, concave function, regarding inequality (3.30), the recent equation implies that:

$$C_p(e) \le c\left(\sum_{j=1}^{|D|}\sum_{i=0}^{h-1} f_p^{(j)}(e,i)\right) \le \sum_{i=0}^{h-1} c\left(\sum_{j=1}^{|D|} f_p^{(j)}(e,i)\right) \tag{3.31}$$

According to equation (3.29), we obtain:

$$\sum_{j=1}^{|D|} f_p^{(j)}(e,i) = |F_p(e,i)|$$

where $F_p(e,i)$ is defined as below:

$$F_p(e,i) = \{v: e \in p_i(v), v \in D\}$$

In Definition 2.5, we defined partition V_i on the vertex set of graph G:

$$V_i = \{\mathcal{L}_{T^s}(u,i): u \in I_i\} \quad \forall i = 0,1,\dots,h$$

In the above equation, we assumed that T^s is an HIT type-1; however, the two other types also partition the vertex set in the same way. We use v_i to partition $F_p(e,i)$ in the following form:

$$F_p(e,i) = \bigcup_{u \in I_i}\{v: e \in p_i(v), v \in D \cap \mathcal{L}_{T^s}(u,i)\}$$

which implies that:

$$\sum_{j=1}^{|D|} f_p^{(i)}(e,i) = |F_p(e,i)| = \sum_{u \in I_i} |\{v: e \in p_i(v), v \in D \cap \mathcal{L}_{T^s}(u,i)\}|$$

Using the recent equation and inequality (3.31), we obtain that:

$$C_p(e) \le \sum_{i=0}^{h-1} c\left(\sum_{u \in I_i}\left|\left\{\begin{matrix} v: e \in p_i(v), \\ v \in D \cap \mathcal{L}_{T^s}(u,i)\end{matrix}\right\}\right|\right) \le \sum_{i=0}^{h-1}\sum_{u \in I_i} c\left(\left|\left\{\begin{matrix} v: e \in p_i(v), \\ v \in D \cap \mathcal{L}_{T^s}(u,i)\end{matrix}\right\}\right|\right)$$

Now we use the definition of the network routing cost function to compute the network routing cost of solution p in cost environment $\mathcal{E}$.

$$C_{\mathcal{E}}(p) = \sum_{e \in E} C_p(e) \le \sum_{e \in E}\sum_{i=0}^{h-1}\sum_{u \in I_i} c\left(\left|\left\{\begin{matrix} v: e \in p_i(v), \\ v \in D \cap \mathcal{L}_{T^s}(u,i)\end{matrix}\right\}\right|\right)$$

Or, equivalently,

$$C_{\mathcal{E}}(p) \le \sum_{i=0}^{h-1}\sum_{u \in I_i}\sum_{e \in E} c\left(\left|\left\{\begin{matrix} v: e \in p_i(v), \\ v \in D \cap \mathcal{L}_{T^s}(u,i)\end{matrix}\right\}\right|\right)$$

In addition,

$$\sum_{e \in E} c\left(\left|\left\{\begin{matrix} v: e \in p_i(v), \\ v \in D \cap \mathcal{L}_{T^s}(u,i)\end{matrix}\right\}\right|\right) = \sum_{e \in p_i(v)} c(|D \cap \mathcal{L}_{T^s}(u,i)|) + \sum_{e \notin p_i(v)} c(0)$$

which implies that ($c(0) = 0$):

$$C_{\mathcal{E}}(p) \le \sum_{i=0}^{h-1}\sum_{u \in I_i}\sum_{e \in p_i(v)} c(|D \cap \mathcal{L}_{T^s}(u,i)|) = \sum_{i=0}^{h-1}\sum_{u \in I_i} |p_i(v)| \cdot c(|D \cap \mathcal{L}_{T^s}(u,i)|)$$

As a result, we find an upper bound of cost $C_{\mathcal{E}}(p)$ for every environment $\mathcal{E} \in \mathbb{E}$ and suggested solution p of the presented scheme in this section:

$$C_{\mathcal{E}}(p) \le \sum_{i=0}^{h-1} \sum_{u \in I_i} \mu_i \cdot c\left(\left|D \cap \mathcal{L}_{T^s}(u, i)\right|\right) \tag{3.32}$$

The following lemma provides a lower bound of the minimum network routing cost in the same problem (cost of optimized solution):

Lemma 3.4 If $\overline{p}^* = (p^*(t_1), p^*(t_2), \ldots, p^*(t_{|D|}))$ denotes an optimal solution of the problem of demand set $D = \{t_1, t_2, \ldots t_{|D|}\}$ in cost environment $\mathcal{E}$,

$$C_{\mathcal{E}}^*(\overline{p}^*) \ge \sum_{u \in J_i - \{s\}} \phi_i \cdot c\left(\left|D \cap \mathcal{L}_{T^s}(u, i)\right|\right) \quad \forall i = 0, 1, \ldots, h, \ \mathcal{E} \in \mathbb{E} \tag{3.33}$$

such that T^s is some HIT in G corresponding to $\overline{I}_s$, and for every $i = 0, 1, \ldots, h$, $J_i \subseteq I_i$ is a $(2\delta_i + 2\phi_i + 1)$ independent set.

Proof. Since $\mathcal{E} \in \mathbb{E}$, the network routing cost of the optimal solution is:

$$C_{\mathcal{E}}^*(\overline{p}^*) = \sum_{e \in E} c\left(f_{\overline{p}^*}(e)\right)$$

where $f_{\overline{p}^*}(e) = \sum_{j=1}^{|D|} f_{\overline{p}^*}^{(j)}(e)$.

Moreover, if edge e does not belong to $p^*(t)$ for every $t \in D$, e will have no flow and its routing cost will be zero. Consequently, we obtain the following equation:

$$C_{\mathcal{E}}^*(\overline{p}^*) = \sum_{e \in \bigcup_{t \in D} p^*(t)} c\left(f_{\overline{p}^*}(e)\right)$$

Now, in order to find a lower bound for $C_{\mathcal{E}}^*(\overline{p}^*)$, we only consider the cost of routing to those demand vertices that belong to $\mathcal{L}_{T^s}(u, i)$ for some $u \in J_i - \{s\}$. In fact, we only compute the routing cost of those paths in the form $p^*(v)$ such that:

$$v \in \bigcup_{u \in J_i - \{s\}} D \cap \mathcal{L}_{T^s}(u, i) \tag{3.34}$$

Regarding relation (3.34), v also belongs to $\mathcal{L}_{T^s}(u, i)$ (for some $u \in J_i - \{s\}$), which implies that $|p^*(v)| \ge d_G(s, v) \ge \phi_i$, as shown in equation (3.24). As a result, we can divide each $p^*(v)$ into two subpaths $p_1^*(v)$ and $p_2^*(v)$ such that $p_1^*(v)$ starts at vertex v, $p_2^*(v)$ ends at s, and $|p_1^*(v)| = \phi_i$. By ignoring routing cost of $p_2^*(v)$, we obtain:

$$C_{\mathcal{E}}^*(\overline{p}^*) \ge \sum_{e \in P} c\left(f_{\overline{p}^*}(e)\right) \tag{3.35}$$

where $P = \bigcup_{u \in J_i - \{s\}} P_u$ and $P_u = \bigcup_{v \in D \cap \mathcal{L}_{T^s}(u, i)} p_1^*(v)$.

Considering u_1 and u_2 as two different members of J_i, we will show that $P_{u_1} \cap P_{u_2} = \emptyset$. By contradiction, assume that there is some edge in both P_{u_1} and P_{u_2}. As a result, there are two vertices v_1 and v_2 such that $v_1 \in D \cap \mathcal{L}_{T^s}(u_1, i)$, $v_2 \in D \cap \mathcal{L}_{T^s}(u_2, i)$, and $p_1^*(v_1) \cap p_1^*(v_2) \ne \emptyset$. Let w denote the vertex in which $p_1^*(v_1)$ and $p_1^*(v_2)$ meet each other. Moreover, assume that q_1 is the shortest path in G from u_1 to v_1 (q_4 is defined similarly for u_2 and v_2). See figure 3.6, which specifies paths q_1, q_2, q_3, and q_4 (as you can see, q_2 represents the subpath of $p_1^*(v_1)$ connecting v_1 to w, and q_3 is between v_2 and w). Since v_1 is in some $\mathcal{L}_{T^s}(u, i)$ such that $u \ne s$, regarding equation (3.22), $|q_1| = d_G(u_1,$

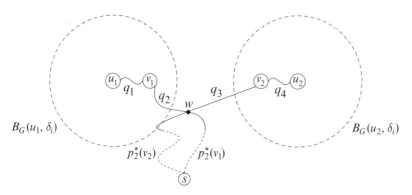

Figure 3.6 Solution paths $p\star(v_1)$ and $p\star(v_2)$ cross each other at w. v_1 and v_2 belong to $\mathcal{L}_{T^s}(u_1, i)$ and $\mathcal{L}_{T^s}(u_2, i)$, respectively.

$v_1) \leq \delta_i$. For the same reason, $|q_4| = d_G(u_2, v_2) \leq \delta_i$. Moreover, as $\left| p_1^\star(v_1) \right|$ and $\left| p_1^\star(v_2) \right|$ are not greater than ϕ_i, $|q_2| \leq \phi_i$ and $|q_3| \leq \phi_i$. Finally, we conclude that path $q_1 \cup q_2 \cup q_3 \cup q_4$ is a path of length not greater than $2\phi_i + 2\delta_i$ from u_1 to u_2, which contradicts the assumption that J_i is a $(2\delta_i + 2\phi_i + 1)$ independent set.

Concerning the two different vertices u_1 and u_2 in $J_i - \{s\}$, $P_{u_1} \cap P_{u_2} = \varnothing$, we rewrite inequality (3.35) in the following form:

$$C_{\mathcal{E}}^\star(\vec{p}^\star) \geq \sum_{u \in J_i - \{s\}} \sum_{e \in P_u} c\left(f_{\vec{p}^\star}(e)\right) \qquad (3.36)$$

Assume that for every $v \in D \cap \mathcal{L}_{T^s}(u, i)$, path $p\star(v)$ is equal to some specific path p such that $|p| = \phi_i$. In this case, as for every edge in p, $f_{\vec{p}^\star}(e) = |D \cap \mathcal{L}_{T^s}(u, i)|$, we obtain that:

$$\sum_{e \in P_u} c\left(f_{\vec{p}^\star}(e)\right) = \phi_i \cdot c\left(|D \cap \mathcal{L}_{T^s}(u, i)|\right)$$

Moreover, since c is a concave function, cost of this case is a lower bound of the incurred cost in other possible cases. Subsequently,

$$\sum_{e \in P_u} c\left(f_{\vec{p}^\star}(e)\right) \geq \phi_i \cdot c\left(|D \cap \mathcal{L}_{T^s}(u, i)|\right) \quad \forall u \in J_i - \{s\} \qquad (3.37)$$

Inequalities (3.36) and (3.37) imply that the proof is complete. ∎

Graph $G_i(I_i, E_i)$ is called the *level-i graph of* G if:

$$\forall u, v \in I_i: (\{u, v\} \in E_i) \leftrightarrow (d_G(u, v) \leq 2\delta_i + 2\phi_i)$$

Lemma 3.5   If $\chi(G_i)$ denotes the chromatic number[4] of graph G_i, the following condition holds:

4. The smallest number of colors needed to color graph $G = (V, E)$ in a way that for every edge $\{u, v\} \in E$, u and v have different colors. Chromatic number of G is denoted by $\chi(G)$.

$$\exists J_i: \sum_{u \in I_i} c\big(|D \cap \mathcal{L}_{T^s}(u, i)|\big) \le \chi(G_i) \cdot \sum_{u \in J_i - \{s\}} c\big(|D \cap \mathcal{L}_{T^s}(u, i)|\big) \qquad (3.38)$$

such that T^s is some HIT in G corresponding to $\bar{I}_s$, and for every $i = 0, 1, \ldots,$ h, $J_i \subseteq I_i$ is a $(2\delta_i + 2\phi_i + 1)$ independent set.

In the exercises, you will be asked to prove Lemma 3.5. Finally, we obtain an upper bound for the competitive ratio of routing scheme $\mathbb{S}_{T^s}$:

Theorem 3.4 The competitive ratio of the versatile routing scheme $\mathbb{S}_{T^s}$ for the mentioned set of possible cost environments $\mathbb{E}$ has the following upper bound:

$$\mathrm{CR}(\mathbb{S}_{T^s}, \mathbb{E}) \le h \cdot \max\left\{ \frac{\chi(G_i) \cdot \mu_i}{\phi_i} : i = 0, 1, \ldots, h-1 \right\} \qquad (3.39)$$

Proof. The theorem is directly obtained by Inequalities (3.32), (3.33), and (3.38). ∎

3.3   Summary and Outlook

In this chapter, we introduced two different approaches to construct the routing scheme for the oblivious routing problems. In both approaches, we tried to make a hierarchical decomposition and then compute a path that connects the successive levels of the graph hierarchical decomposition together.

In the first section, we described a top-down versatile routing scheme proposed by Gupta et al. (2006). We explained that how the scheme is implemented using the hierarchical decomposition tree mentioned in chapter 2. Additionally, we mathematically analyzed the competitive ratio of both the fractional and integral types of the scheme in some specific range of the routing cost environments.

In the second section, a bottom-up versatile routing scheme for unweighted graphs was introduced that is based on the hierarchical independence trees defined previously. Moreover, we analytically discussed on how the competitive ratio of the presented scheme can be bounded to a deterministic value.

Exercises

1. During the proof of Theorem 3.2, we presented an algorithm which constructed a weighted graph of unit weight function for every given weighted graph. Show that the cost incurred by any solution of an oblivious routing problem in weighted graph G is not less than the cost of the corresponding solution in the graph with unit weight function (G').

2. Prove Claim 3.1.
3. Prove equation (3.17).
4. Regarding Definition 3.5, prove that for every closed cut set X of the ith commodity of a general routing problem of the routing cost environment E and graph $G = (V, E, w)$, the following equation holds:

$$\sum_{e \in X} f_{\bar{F}}^{(i)}(e) = \Pi_3(K_i)$$

where $\bar{F}$ denotes a fractional solution of the general routing problem in environment $\mathcal{E}$.
5. Prove Lemma 3.5.
6. The *doubling-dimension* of graph $G = (V, E)$ is defined as the smallest ρ such that for every even number r and vertex $v \in V$, there exist at most 2^ρ balls $B_1, B_2, \ldots, B_{2^\rho}$ of radius $r/2$ in graph G such that:

$$B_G(v, r) \subseteq \bigcup_{i=1}^{2^\rho} B_i$$

Assuming that graph G has the doubling dimension of ρ, simplify the inequality obtained by Theorem 3.4 (*Hint*: Find an upper bound for the value of $\chi(G_i)$ for every $i = 0, 1, \ldots, h - 1$).
7. Briefly describe the routing scheme presented by Srinivasagopalan, Busch, and Iyengar in 2012. What are the properties of the routing scheme that makes it distinguished?
8. Briefly describe the routing scheme presented by Gupta et al. (2006). What are the properties of the routing scheme that distinguish it?

Suggested Reading

Bartal, Y. 1996. Probabilistic approximation of metric spaces and its algorithmic applications. *Proceedings of the 37th FOCS*, 184–193.

Chuzhoy, J., A. Gupta, J. S. Naor, and A. Sinha. 2008. On the approximability of some network design problems. *ACM Transactions on Algorithms*, 4(2): 1–17.

Fakcharoenphol, J., S. B. Rao, and K. Talwar. 2003. A tight bound on approximating arbitrary metrics by tree metrics. *Proceedings of the 35th STOC*, 448–455.

Fraigniaud, P., E. Lebhar, and L. Viennot. 2007. The inframetric model for the Internet, Tech. Rep., INRIA Paris-Rocquencourt.

Garg, N., R. Khandekar, G. Konjevod, R. Ravi, F. S. Salman, and A. Sinha. 2001. On the integrality gap of a natural formulation of the single-sink buy-at-bulk network design formulation. *Proceedings of the 8th IPCO*, 170–184.

Goel, A., and I. Post. 2009. An oblivious $O(1)$-approximation for single source buy-at-bulk. Annual IEEE Symposium on Foundations of Computer Science, 442–450.

Guha, S., A. Meyerson, and K. Munagala. 2000. Hierarchical placement and network design problems. *Proceedings of the 41st FOCS*, 603–612.

Gupta, A., M. T. Hajiaghayi, and H. Racke. 2006. Oblivious network design. *Proceedings of the Seventeenth Annual ACM-SIAM Symposium on Discrete Algorithms, SODA '06*. New York: ACM, 970–979.

Gupta, A., R. Krauthgamer, and J. R. Lee. 2003. Bounded geometries, fractals, and low-distortion embeddings. *Proceedings of the 44th FOCS*, 534–543.

Gupta, A., M. Pal, R. Ravi, and A. Sinha. 2004. Boosted sampling: Approximation algorithms for stochastic optimization problems. *Proceedings of the 36th STOC*, 417–426.

Harrelson, C., K. Hildrum, and S. B. Rao. 2003. A polynomial-time tree decomposition to minimize congestion. *Proceedings of the 15th SPAA*, 34–43.

Konjevod, G., A. W. Richa, and D. Xia. 2008. Dynamic routing and location services in metrics of low doubling dimension. *Proceedings of DISC '08*, 379–393.

Kuhn, F., T. Moscibroda, and R. Wattenhofer. 2005. On the locality of bounded growth. *Proceedings of PODC '05*. New York: ACM, 60–68.

Rake, H. 2002. Minimizing congestion in general networks. *Proceedings of the 43rd FOCS*, 43–52.

Salman, F. S., J. Cheriyan, R. Ravi, and S. Subramanian. 2000. Approximating the single-sink link-installation problem in network design. *SIAM Journal on Optimization*, 11(3):595–610.

Srinivasagopalan, S. 2011. Oblivious buy-at-bulk network design algorithms. Doctoral dissertation, Louisiana State University, Baton Rouge, LA.

Srinivasagopalan, S., C. Busch, and S. S. Iyengar. 2009. Brief announcement: Universal data aggregation trees for sensor networks in low doubling metrics. *Algorithmic aspects of wireless sensor networks, Proceedings of the 5th International Workshop, ALGO-SENSORS*. Berlin: Springer-Verlag, 151–152.

Srinivasagopalan, S., C. Busch, and S. S. Iyengar. 2011. Oblivious buy-at-bulk in planar graphs. *Workshop on Algorithmic Computing, WALCOM*, IIT-Delhi, 18–20 February.

Srinivasagopalan, S., C. Busch, and S. S. Iyengar. 2012. An oblivious spanning tree for single-sink buy-at-bulk in low doubling-dimension graphs. *IEEE Transactions on Computers*, 61(5):700–712.

Talwar, K. 2002. Single-sink buy-at-bulk LP has constant integrality gap. *Proceedings of the 9th IPCO*, 475–486.

II Applications

Chapter 4   A Secure Versatile Model of Content-Centric Networks

Contemporary Internet architecture is based on the host-based model whereby the data flow is driven by explicit addresses assigned to the communicating hosts. However, in the future, Internet structure is expected to be focused on content and not the physical location of the machines. This leads to a very efficient use of a network in which many people are interested in the same content. In other words, content-centric networking seeks to adapt Internet architecture to the current usage pattern.

One of the most critical and basic issues in designing content-centric networks is the security of the system. Considering the data network as a service, users need assurance regarding security and privacy. In today's Internet, preservation of security is already implemented in different layers including IP, TCP, and the application layer. However, there are many unsolved challenges regarding security issues in a host-oblivious network as the identity of each host may be unknown. For content-centric networks as a network of oblivious hosts, it is necessary for network designers to come up with an appropriate security scheme that is not based on the IPs or locations of the hosts.

The other important challenge in the data distribution systems is how to distribute the data flow throughout the network. More specifically, in a content-centric network, some of the hosts get congested because the data traffic pattern has a lot of data flowing through them. As an illustration, consider a situation that some special content goes viral over a short period of time. In this case, the hosts that are in any way related to that content and their neighbors may get congested with a huge flow of data. As a result, network delay and energy consumption may increase substantially.

Regarding the two aforementioned issues in content-centric networks, we are proposing a model of a data distribution system that is a peer-to peer network and deals with both issues appropriately. In this model, the data requests and the level of security are managed in a content-centric way;

however, the data flows are routed in a node-congestion prevention routing scheme that is location/host based. We call this the hybrid model.

In section 4.1, we present an introduction of the security issues in content-centric networks. Then, the details of our hybrid model in the context of its security and routing schemes will be addressed in sections 4.2 and 4.3. In sections 4.4 and 4.5, we explain how our routing scheme prevents node congestion and at the same time does not increase the routing cost. In order to prove our claim, we provide some mathematical and geometrical analysis in sections 4.6 and 4.7.

4.1   Security Preliminaries

The first step of understanding the information security is to understand its basic principles. *Confidentiality*, *integrity*, and *availability* (CIA) comprise all the principles on which every security program is based. Other key concepts have been proposed, such as accounting, auditing, and non-reputation, but they are not as essential as CIA.

Confidentiality   Confidentiality determines the secrecy of the information asset. It refers to preventing the disclosure of some information to unauthorized individuals or systems. For instance, a credit card transaction over the Internet requires the card number to be transmitted from the purchaser to the merchant and from the merchant to a transaction-processing network. The system attempts to enforce confidentiality by encrypting the card number during the transmission. This is done by limiting the places where it might appear (in databases, log files, backups, printed receipts, and so on), and also by restricting access to the places where it is stored. If an unauthorized party obtains the card number in some way, a breach of confidentiality has occurred.

Confidentiality is necessary for maintaining the privacy of the people whose personal information is held by a system. Authentication methods such as user IDs and passwords (which uniquely identify the users and control data access to the system resources) underpin the goal of confidentiality.

Integrity   Integrity refers to the trustworthiness of information resources. In information security, data integrity means maintaining and ensuring the accuracy and consistency of data over its entire life cycle. In fact, the data cannot be modified in an unauthorized or undetected manner. This is not the same as what is called referential integrity in the databases; however, it can be viewed as a special case of consistency as understood in the classic model of transaction processing. Integrity is violated when a message is actively modified in transit. Information security systems typically provide message integrity in addition to data confidentiality.

Availability Availability refers to the condition in which the information resources are available. In any information system, the information must be available when it is needed; otherwise, the system may fail to reach its goal. This means that the computing systems used to store and process the information, the security controls used to protect it, and the communication channels used to access it must be functioning correctly. High-availability systems aim to remain available at all times and prevent service disruptions due to power outages, hardware failures, and system upgrades. Ensuring availability also involves the prevention of the denial-of-service (DoS) attacks.

Privacy Privacy relates to all the elements of the CIA triad. It considers which information can be shared with others (confidentiality), how that information can be accessed safely (integrity), and how it can be accessed (availability).

Host-Oblivious Security Schemes Ensuring security and privacy in a host oblivious network is a challenging problem, especially in a content-based network. We distinguish a host-oblivious network security paradigm from a conventional host-dependent network security paradigm from two perspectives, a host-oblivious security association and multiple security levels.

Host-dependent network security is based on the conventional networks themselves, which run a host-based architecture. A source initiates communication on the assumption that the source is able to identify a destination. Due to the destination awareness, for cryptographic functions, the source simply uses a shared symmetric secret key or a public key of the destination. Because the destination can also identify the source, a security association for both nodes is easily established. That is, regardless of the diversity of the application data, only a single security level is provided during the entire communication (which is based on the hosts' identities).

There are two basic principles of the host-oblivious network security. The first is that a security association is independent of the host identification, which is caused by blindness of hosts in content-based network architecture. The second is the provision of multiple security associations for diverse security sensitivity of the various contents.

An important challenge in oblivious content-centric network security is the protocol design to bootstrap the establishment of secure communication infrastructure from a collection of nodes that may have been pre-initialized with some secret information without any prior direct contact among them. This problem is called the *bootstrapping problem*. The key point in host oblivious security is the selection of the keys, which must be independent of each other and of the host. As mentioned previously, there are two methods for generating session keys, including symmetric and asymmetric crypto-algorithms. The prerequisite for applying public-key cryptography is the awareness of the destination address, which is a contradiction with content-centric networks.

There are three types of general key agreement schemes: the trusted-server scheme, the self-enforcing scheme, and the key predistribution scheme. In key predistribution, the key information is distributed among all nodes prior to deployment. The random pool-based (RPB) scheme is a proposed random key predistribution scheme designed to address the bootstrapping problem. This method in brief is as follows: A random pool of keys is selected from the key space; then, each node receives a random subset of keys from the key pool prior to any connection that it wants to be involved in. Any node that has a common key within its subset can apply it as a shared secret key to initiate communication.

There is a key distribution center (KDC) in a random pool-based scheme that manages a key pool. Before deploying the nodes, an initialization phase is performed. In the initialization phase, the basic scheme picks a random pool (set) of keys S out of the total possible key space. For each node, m keys are randomly chosen from the key pool S; then, KDC distributes the selected keys to the nodes.

When the nodes are deployed, a key setup phase is performed. When two nodes want to generate a link key, they exchange the keys' indexes. In the case that two nodes share at least one key, they use one of the commonly shared keys as a *link key*. This method has been extended to the *q-composite random pool based scheme*. In this scheme, both nodes are required to share at least q numbers of keys and the link key is generated by combining q common keys.

4.2   The Hybrid Model Description

In this section, we propose a hybrid model of the content-centric networks that will be analyzed in the context of security and cost efficiency through the chapter.

We assume that there are a number of *cyber devices* scattered over a plane. These devices, which may have time-varying locations (such as mobile devices) have various interests with respect to special data contents. Additionally, each device has some stored data that may satisfy others' interests. The data transmission between the cyber devices is performed through a network of *routing nodes*, which are deployed on a convex subset of the Euclidean plane and are wired together. Let $r_1, r_2, ..., r_n$ denote the n routing nodes in the network. Despite the cyber devices, every routing node has a fixed location on the conveying convex subset.

Moreover, we assume that if a cyber device gets close enough to some routing node, a wired or wireless connection will be established between them; however, every device can only be connected to one routing node at a given time. Note that the cyber devices may get connected to and disconnected from the network of the routing nodes as they are moving through the plane over time. More specifically, each cyber device may get connected to different routing nodes as it moves in the plane; also, it may move far from the network such that it gets connected to no routing node. See figure 4.1 for an illustration.

Now consider the following scenario: Some cyber device is interested in receiving some data, say a movie. We call this device the interested

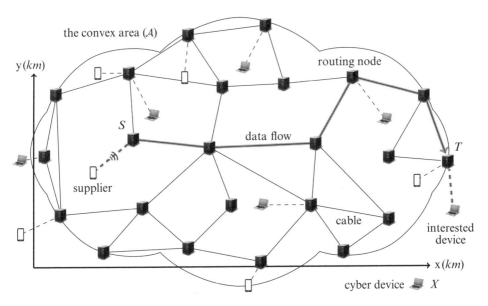

Figure 4.1 Schematic representation of the network topology and node deployment in the Euclidean plane. Note that device X is not connected to any routing node.

device. Assuming that this device is close enough to some routing node (T), a *request* including the movie name is sent from the interested device to T via a wired or wireless connection between them. Then, node T will broadcast the request through the network of routing nodes in the following way: "Each routing node that receives the request, will forward it to its connected cyber devices and neighboring routing nodes." In the case that there exists a connected cyber device that has the requested movie, it notifies the interested device with a message through the reverse path. When the notification messages are delivered to the interested device, one of the cyber devices that has replied to the request will be chosen as the supplier (for simplicity, we assume that the sender of the first received notification will be chosen as the supplier). Then, the interested device asks the supplier to send the movie. Finally, the supplier transmits the movie through a path between S and T, which is obtained by a versatile routing scheme (note that the data has to be transmitted from the supplier to S in the first step and from T to the interested device in the last step).

In this model, we assume that every cyber device wants to satisfy its demand only by receiving the content of interest from an authenticated device. In addition, the supplier will share its data only with the authenticated devices. To deal with these security issues, we use a security paradigm that uses a trusted third party to provide such authentication between devices. Furthermore, we will thoroughly address the cost efficiency and the node-congestion issues regarding the data flows through the network of our model.

Now we will look at the detailed four phases of the protocol that is used in this model and is mainly based on the security paradigm proposed by Chan et al. (2003.) and Jeong et al. (2010). In this protocol, we assume that every cyber device connected to the network has a secure connection with a trusted third party in the network called *the trusted node*. This node has a *key pool* that contains a number of keys that are used to establish authenticated connections in the network. Additionally, we assume that the routing nodes are already able to route data through the network using a location-based routing scheme, i.e., in the case that the final destination of an arrived packet is specified, the routing node will forward the packet toward a suitable interface. We specify the destination of a packet using the ID of the routing node in which the packet is supposed to be in the last step.

Phase 0: Key Distribution   When a cyber device gets connected to the network of the routing nodes, it asks the trusted node for m keys. Then, the trusted node assigns m keys of the key pool to the cyber device as its key ring. This is done in such a way that for every given pair of connected devices x and y, device x authenticates y if and only if a specific q-size subset of x's key ring completely belongs to the key ring of y ($q \leq m$).

Here is the exact format of the "registration message" which is sent by the trusted node using the location-based routing scheme and includes the key ring of the new connected device:

$$REG = \left(RNI, \{k_1, k_2, \dots, k_m\}_{pvt\text{-}key}\right) \qquad (4.1)$$

where RNI is the index of the routing node directly connected to the new device (i.e., if the new device is connected to r_3, RNI = 3). Also, keys $k_1, k_2, \dots, k_m$ belong to the key pool and are assigned to x as its key ring. These keys are encrypted in the trusted node before sending the REG message. The encryption of the key ring is done by the private key of the trusted node (pvt-key) using a symmetric key-based authentication algorithm.

Phase 1: Requesting Content   The interested device makes a "request message" containing a unique ID of the request, the identifying information of the exact content in which it is interested, a TTL[1] integer value, and the indexes of the q randomly chosen keys belonging to its key ring ($q \leq m$); for example, the device wants to receive content c, TTL = 15, and it chooses the keys of indexes $i_1, i_2, \dots, i_q$ among the m keys belonging to its key ring ($\forall j = 1, 2, \dots, q\colon 1 \leq i_j \leq m$). Then, the message will be sent to the routing node directly connected to the device. This node will broadcast the request message through the network by forwarding it to its connected devices and neighboring routing nodes. Before forwarding the message, the routing node decreases the TTL by one. Every other routing node that receives the request message also decrements its TTL and broadcasts it through the network in the same way until the TLL value becomes zero.

The detailed content of the request message is as follows:

$$REQ = \left(RID, \{CID\}_k, TTL, (i_1, i_2, \dots, i_q)\right) \qquad (4.2)$$

where RID and CID denote the request ID and the content ID, respectively, and uniquely specify the request and the content. Additionally, in order to preserve the privacy of the interested device, value CID is included in the request message after getting encrypted by key k. This key is obtained by applying a globally known hash function (h) to the q chosen keys of the key ring, i.e., considering the key ring as the set of keys in the form $\{k_i\colon i = 1, 2, \dots, m\}$, the key value is computed by the following formula:

$$k = h\left(k_{i_1}, k_{i_2}, \dots, k_{i_q}\right) \qquad (4.3)$$

---

1. Time to live.

Responding to the Request Every cyber device that receives the request message checks whether its key ring includes the q keys specified in the message. If yes, it then checks its own data storage to see if it has the requested content. In the case that it does not have all the q keys or the exact requested content, the request message will simply be ignored; otherwise, the device makes a "notification message" containing the index of the routing node to which it is already connected, the TTL value of its associated request message (represented by RES[2]), and the ID of the request. Then, this message will be sent to the interested device using the reverse path of the one through which the corresponding request message was previously sent (later, we will address the details of how the reverse path is computed).

The detailed content of the notification message is as follows:

$$\text{NOTIF} = (\text{RID}, \text{RNI}, \text{RES}) \tag{4.4}$$

where RNI is the index of the routing node directly connected to the device that is sending the notification message. Note that the value of RID is the same as what was included in the corresponding REQ message.

Phase 3: Choosing the Supplying Device When all of the notification messages are delivered to the interested device, it will choose the one with the largest RES value, i.e., the message with the closest sender. The sender of the chosen message is called the *supplier*. Then, the interested device sends an acknowledgment message to the supplier to ask for content c. The acknowledgment message includes the ID of the routing node connected to the interested device and is sent through a path obtained by the location-based routing scheme. The detail of the acknowledgment message is as follows:

$$\text{ACK} = (\text{RID}, \text{RNI}_I, \text{RNI}_S) \tag{4.5}$$

where RNI_I and RNI_S denote the indexes of the routing nodes directly connected to the interested device and the supplier, respectively. Again, note that the value of RID is the request ID, which was put in the corresponding REQ and NOTIF messages.

Phase 4: Data Transmission When the supplier receives the acknowledgment message, it starts sending the requested content to the interested device through a congestion prevention routing scheme, which will be discussed in detail later. The data transmission occurring in this phase is usually much larger than the messages transmitted previously. Henceforth, a

2. Residue.

number of data messages will be sent in this phase in such a way that each message contains only a portion of the whole set of data that has to be transmitted toward the interested device. The parameters of a data message are as follows:

$$\text{DATA} = \left(\text{RID}, \text{RNI}_I, \text{RNI}_M, \{\text{FIRST}, \text{LAST}, \text{DP}\}_{k'}\right) \qquad (4.6)$$

where FIRST and LAST denote the indexes of the first and the last bytes of the data portion which is included in the data message, respectively. Additionally, DP represents the data portion of size LAST − FIRST, and value RNI_M specifies the index of the intermediate routing node through which the data message will be routed toward the interested device (in section 4.4, we will describe the way of choosing the intermediate routing node in more detail). Note that in equation (4.6), the data included in the message is encrypted using key k', which is obtained by the following equation:

$$k' = h_{\text{content-type}}\left(k_{i_1}, k_{i_2}, \dots, k_{i_q}\right) \qquad (4.7)$$

such that the hash function $h_{\text{content-type}}$ depends on the type of the requested data content. For example, in the case that the interested device requests a multimedia file, we use a hash function that generates a short key; however, when the interested device requests a file that urges a higher level of security, we have to use another hash function, which generates longer keys.

When the data message is delivered to the interested device, it is decrypted using key k' computed by equation (4.7) in the interested device.

4.3   Message Forwarding in the Routing Nodes

In the previous section, we described the hybrid protocol for distribution of interesting data through the proposed network model. Additionally, different types of messages used in this protocol were precisely determined. In the present section, we will focus on the implementation of message forwarding in the routing nodes. We also illustrate the way that the cyber devices communicate with each other using a state machine diagram.

In this section, we assume that for a pair of connected routing nodes X and Y, node X needs to have a dedicated interface to handle its connection with Y and vice versa. Additionally, every cyber device connects to a routing node through a "port" of the node.

Forwarding REQ and NOTIF   As mentioned before, the message REQ is broadcasted through the network in Phase 1. At the first hop, the interested

v	RID	Interface #
1	111	11
0	001	10
1	011	10
1	101	00

Broadcast Table of node r

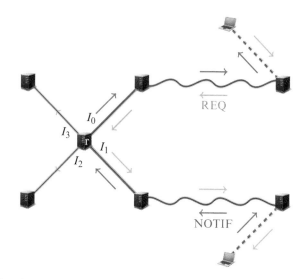

Figure 4.2 How routing node r forwards the REQ and NOTIF messages using its broadcast table. Note that node r has four interfaces.

device sends the message to the pth port of its directly connected routing node T. Node T stores the message RID and port index p in a table before forwarding the message toward other devices and the neighboring routing nodes. During the process of broadcasting the message, every routing node that receives the message via its ith interface will forward it to the routing nodes connected to other interfaces (other than the ith one). The routing node stores the RID (request ID) of the message REQ and the index (i) of the interface through which the message was received in the broadcast table, which is depicted in figure 4.2.

The forwarding process of a NOTIF message is the same as what is done for its associated REQ message, but in the reverse order. This is because the NOTIF message has to be routed in the reverse path of the one through which the request message was sent. In order to route through the reverse path, every routing node that receives the NOTIF message searches through the broadcast table to find the RID of the message. Then, it will forward the message through the corresponding interface of the RID in the table (see figure 4.2 for illustration). This process continues until the NOTIF message reaches routing node T, which is directly connected to the interested device. In this step, there is no valid entry associated with the message RID in the broadcast table of node T; however, node T finds the message RID in another table where the corresponding port index of the interested device has already been stored. Finally, the message gets delivered to the interested device via the appropriate port.

Note that the first routing node that receives the NOTIF via its jth port stores the message RID and the value j in a table so that it can remember the port index if a response message (ACK) comes into it in the future.

Forwarding REG and ACK   As mentioned before, we assume that REG and ACK are routed by a location-based routing scheme. This scheme is implemented using a number of tables known as forwarding tables. Every routing node has a forwarding table that specifies the outgoing interface of any incoming message based on its destination.

Messages REG and ACK both have the indexes of the routing nodes directly connected to their target devices (RNI in REG and RNI_I in ACK). Henceforth, every routing node that receives a registration message searches through the forwarding table to specify the outgoing interface corresponding to the RNI value of the message and then forwards it through the specified interface. Finally, when the routing node of index RNI receives the REG message, it will deliver it to the recently connected device. An ACK message is forwarded in a similar way. However, when the message reaches its final routing node, it is forwarded based on the table that has already stored the port index associated with the message RID.

Forwarding DATA   In a data message, there are two destinations: the intermediate routing node and the ultimate node. The index of the earlier node is specified by RNI_M, and the latter node is determined by index RNI_I.

When a routing node gets a DATA message, it follows algorithm 4.1 to forward the message. In this algorithm, forwarding the DATA message to a routing node is assumed to be done based on the aforementioned location-based routing scheme; however, the message will be forwarded to the target device based on the table that has already stored the port index associated with the RID value of DATA.

Finally, in this section, we provide a detailed interaction of every cyber device in the proposed protocol. Figure 4.3 specifies a state machine that illustrates how a connected device communicates with other elements of the network during different phases of the protocol.

As shown in figure 4.3, when a cyber device gets connected to the network, it asks the trusted node for m keys and goes to the start state. Then, it transits to the ready state if it receives an appropriate REG message. If the user asks for content c, the device broadcasts a request message, sets the timer to constant $\mathcal{I}$, and goes to the demand state. Additionally, if the

---

Algorithm 4.1   DATA forwarding

Input: DATA message
1    $k \leftarrow$ the routing node index;
2   **if**   DATA.RNI$_I = k$ **then**
3 | Forward DATA to the target device;
4 | **return**;
5   **end**
6   **else if**   DATA.RNI$_M = k$ **then**
7 | DATA.RNI$_M \leftarrow$ *null*;
8 | Forward DATA to the routing node of index RNI$_I$;
9   **end**
10   **else if**   DATA.RNI$_M \leftarrow$ *null*   **then**
11 | Forward DATA to the routing node of index RNI$_I$;
12   **end**
13   **else**
14 | Forward DATA to the routing node of index RNI$_M$;
15   **end**

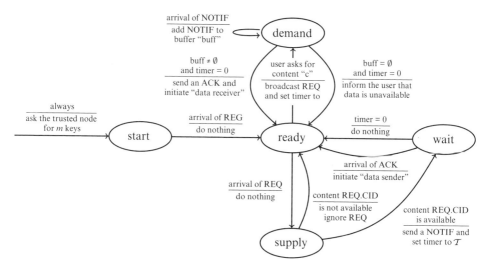

Figure 4.3 How routing node r forwards the REQ and NOTIF messages using its broadcast table. Note that node r has four interfaces.

device is in the ready state and a REQ message arrives at the device, it goes to the supply state.

Now, consider the case that the user is in the demand state. In this case, every arrived NOTIF message will be added to a message buffer called buff. As the timer was set to $\mathcal{I}$ at the moment of state transition from ready to demand, the device stays in the demand state for $\mathcal{I}$ units of time and then it will go to the ready state again. During this state transition, if the buffer is empty, the user gets informed that the requested data is unavailable. However, in the case that the buffer is not empty, the device will send an ACK message and initiate a parallel thread called the data receiver to receive the requested data.

Finally, if the device goes to the supply state, it checks its data storage to see whether it has the content of ID REQ.RID. If it does not have the requested content, it simply ignores the REQ message and goes back to the ready state. However, if it has the content specified in the REQ message, it sends a NOTIF message to the interested device, sets a timer to $\mathcal{I}$, and goes to a new state called wait. If an ACK message arrives at the device within $\mathcal{I}$ units of time after the state transition, the device initiates another parallel thread called data sender, which sends the requested data toward the interested device (in the form of some DATA messages); otherwise, it simply goes back to the ready state without doing anything.

4.4 Oblivious Routing Problem Specification

In this section, we precisely define the oblivious routing problem we encounter in sending data through the hybrid network. To do this, we need to specify the network topology and characteristics using the concept of *geometric* graphs. Preliminary definitions are presented as follows.

4.4.1 Definitions

The triple (V, E, loc) is an unweighted geometric graph in the n-dimensional Euclidean space $\mathcal{S}$ if (V, E) specifies an unweighted graph; and loc denotes a function in the form $loc : V \mapsto \mathcal{S}$.

For every (unweighted) geometric graph, we define three distance types for each pair of vertices. Assuming that u and v are two vertices of graph $G = (V, E, loc)$, the three distance types are as follows.

- The *graph distance* between u and v is denoted $d_G(u, v)$ and defined as $d_{G'}(u, v)$ where $G' = (V, E)$ represents the unweighted graph corresponding to G. Moreover, if p denotes a path in G, its length $len_G(p)$ is defined as $len_{G'}(p)$.

- The (Euclidean) distance between u and v is denoted by $d_{\mathrm{loc}}(u, v)$ and defined as $\|\mathrm{loc}(v) - \mathrm{loc}(u)\|$ (where $\|X - Y\|$ represents the Euclidean distance[3] between points X and Y in space $\mathcal{S}$).
- The *hop-by-hop distance* between u and v is represented by $\delta_G(u, v)$ and defined inductively by the following equation:

$$\delta_G(u, v) = \begin{cases} 0 & u = v \\ \min_{\{w,v\} \in E} \{\delta_G(u, w) + \|loc(w) - loc(v)\|\} & \text{otherwise} \end{cases}$$

Regarding the above defined distances in a geometric graph, it is inferred by the triangular inequality that for every pair of vertices u and v, the Euclidean distance $d_{\mathrm{loc}}(u, v)$ is not greater than $\delta_G(u, v)$. Also, the *pseudo-diameter* Ψ of geometric graph $G = (V, E, \mathrm{loc})$ is defined in the following way:

$$\Psi(G) = \max_{u \in V} \max_{v \in V} \frac{d_G(u, v)}{d_{\mathrm{loc}}(u, v)}$$

For every point $X \in \mathbb{R}^n$ and value $r \in \mathbb{R}_{\geq 0}$, the n-dimensional ball $B(X, r) \subseteq \mathbb{R}^n$ is defined by the following equation:

$$B(X, r) = \left\{ Y \in \mathbb{R}^n : \|X - Y\| \leq r \right\}$$

The n-dimensional space $\mathcal{S}$ is said to be *thoroughly $\mathcal{R}$-covered* by geometric graph $G = (V, E, \mathrm{loc})$ if the following condition holds:

$$\mathcal{S} \subseteq \bigcup_{v \in V} B(\mathrm{loc}(v), \mathcal{R})$$

4.4.2 Graph Representation of the Hybrid Model

As mentioned before, the network of routing nodes consists of n vertices r_1, $r_2, \ldots, r_n$ which are wired together in the form of a connected graph (where r_i represents the ith routing node in our model). Additionally, we assume that the upper bound of the distance between cyber device and its directly connected routing node is positive constant value $\mathcal{R}$; in fact, these devices can get connected to node r_i located in $B(\mathrm{loc}(r_i), \mathcal{R})$.

Let $G = (V, E, \mathrm{loc})$ denote the graph representation of our system where the vertex set V is equal to $\{v_i : i = 1, 2, \ldots, n\}$ and function $\mathrm{loc} : V \mapsto \mathbb{R}^2$ specifies the deployment of r_is in the Euclidean plane. We consider the following constraint for the node locations on the plane:

$$\forall u, v \in V : d_{\mathrm{loc}}(u, v) > R$$

Any geometric graph that holds such constraint is said to have an $\mathcal{R}$-*distant* deployment.

---

3. Euclidean distance between points $X \in \mathcal{S}$ and $Y \in \mathcal{S}$ is denoted by $\|X - Y\|$ and defined as the length of line segment $\overline{XY} : \|X - Y\| = |\overline{XY}|$.

Moreover, we assume that there exists some $c \geq 1$, such that:

$$\forall \{u, v\} \in E : d_{\text{loc}}(u, v) \leq c\mathcal{R}$$

In addition, we assume that there exists some convex subset[4] $\mathcal{A} \subseteq \mathbb{R}^2$ such that:

$$\begin{cases} \text{loc}(v) \in \mathcal{A} \\ \mathcal{A} \subseteq \bigcup_{v \in V} B(\text{loc}(v), \mathcal{R}) \end{cases} \forall v \in V$$

which means that set $\mathcal{A}$ completely contains the network and is *thoroughly* $\mathcal{R}$-covered by the graph representation (G) of the network. This means that if a cyber device is located in area $\mathcal{A}$, there exists a routing node that is not farther than $\mathcal{R}$ from the device and consequently, they can be connected together (see figure 4.4 for illustration).

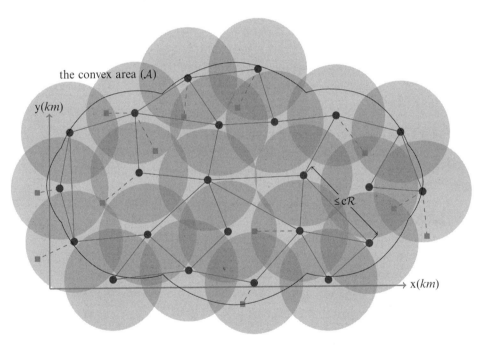

Figure 4.4 Deployment of the routing nodes (gray circles) and cyber devices (gray rectangles) on the Euclidean plane. Note that the radius of any gray circle is $\mathcal{R}$. Note that the routing nodes are $\mathcal{R}$-distant and at the same time, they thoroughly $\mathcal{R}$-cover the convex area $\mathcal{A}$. Additionally, area $\mathcal{A}$ completely includes the network of routing nodes; however, there may exist some cyber devices located outside of $\mathcal{A}$.

4. Set $\mathcal{S} \subseteq \mathbb{R}^n$ is a convex subset of the n-dimensional Euclidean space if for every $X, Y \in \mathcal{S}$, line segment $\overline{XY}$ completely lies inside $\mathcal{S}$.

In order to prove the feasibility of the aforementioned constraints regarding the deployment of vertices in the Euclidean plane, we will show that for any subset S of the Euclidean plane, there is a deployment of nodes in S that is $\mathcal{R}$-distant and thoroughly $\mathcal{R}$-covers S. The following algorithm specifies a way of constructing such a deployment:

In a greedy manner, arbitrarily choose the uncovered point X in set S and fix the node there. Then cover the points belonging to ball $B(X, \mathcal{R})$ (at first, every point in S is uncovered). Do this repetitively until there remains no uncovered point in S.

It is already clear that this greedy algorithm makes a thoroughly $\mathcal{R}$-covered node deployment in S. Therefore, we only need to show that the generated deployment by the above algorithm is $\mathcal{R}$-distant. By contradiction, assume that it is not, i.e., there are two nodes located at points $X \in S$ and $Y \in S$ such that the value $|\overline{XY}|$ is not greater than $\mathcal{R}$; or equivalently, Y is in the ball $B(X, \mathcal{R})$. Without loss of generality, assume that X is chosen earlier than Y in the greedy algorithm. Thus, when we chose Y as a node location, all the points of S inside the ball $B(X, \mathcal{R})$ (including Y) had been covered earlier. This means that when we chose Y in the algorithm, it had already been covered; however, according to the algorithm, each node location should be uncovered at the time of being chosen.

4.4.3   Oblivious Routing Cost Environment

Note that in the proposed protocol of our model, we assume that any cyber device can ask for data in some specific moment, and its request may be satisfied by other devices in some time interval. Additionally, we have assumed that our cyber devices may move through the network and get connected to different routing nodes. These assumptions make the distribution of the data traffic pattern in our model dynamic and time-sensitive.

Here, we consider the data flow of our model in ϵ-length time intervals (for some small $\epsilon > 0$). Assume that in interval $I_x = [x, x + \epsilon]$, there are $k(x)$ data flows through the network in such a way that routing node $s_i(x)$ sends $b_i(x)$ bits of data to node $t_i(x)$ for every $i = 1, 2, \ldots, k(x)$ (note that $s_i(x)$ and $t_i(x)$ can be any two vertices of graph G and $b_i(x)$ is an arbitrary positive number). Henceforth, in order to find the data paths in time interval I_x, we need to solve a general routing problem of commodities $\bar{K} = (K_1(x), K_2(x), \ldots, K_{k(x)}(x))$ in graph G such that:

$$K_i(x) = (s_i(x), t_i(x), b_i(x)) \quad \forall i \in k(x)$$

Note that the sequence of commodities is a function of time x and may change over time. To illustrate, assume that one device always asks for a huge bunch of data, which leads to traffic congestion in the routing node connected to the device. As the device moves through the network over time, it gets connected to different routing nodes, which leads to changes in the target of some commodities.

For the balance of this chapter, we will focus on data routing through the network while satisfying the following concerns:

1. Preventing data congestion in some specific routing node (*vertex-level* cost optimization): Data congestion increases the delay and energy consumption of our network. Moreover, as the cameras are power consumers, distributing the electrical load through the network will improve the energy efficiency.

2. Keeping the total number of data sending/forwarding small (*network-level* cost optimization) as the total energy consumption of our system is substantially dependent on the amount of data transmitted between the routing nodes.

If our goal is to lessen node traffic congestion, as mentioned in chapter 1, we have to consider the edge routing cost function as the summation $(\text{cost}_e = \Sigma_f)$; and also the nrc_π will be in the following form (see equation 5.1):

$$\text{nrc}_\pi\big(c(e_1), c(e_2), \ldots, c(e_{|E|})\big) = \frac{1}{2}\max_{v \in V} \sum_{\substack{e \in E \\ v \in e}} c(e)$$

Consequently, we have to deal with a general routing problem in a time-varying routing cost environment in the form $\mathcal{E}(x) = \left(\sum, 1/2\max_v \sum_{e'}, \bar{K}(x)\right)$. This implies that the general routing problem turns into an oblivious one when we consider a long time interval containing multiple ϵ-length intervals (note that for every time interval I_x, we may have a different unpredictable sequence of commodities $\bar{K}(x)$). Henceforth, the set of all the possible routing cost environments will be in the following form:

$$\mathbb{E} = \left\{ \left(\sum, \frac{1}{2}\max_v \sum_{e'}, \mathcal{K} \right) : \mathcal{K} \in \mathcal{K} \right\} \tag{4.8}$$

where $\mathcal{K}$ is the set of the commodity sequences which may contain any number of commodities and each of the commodities may have any source, target, and value.

Additionally, in order to address the second mentioned concern, we need to route data in such a way that the following value does not exceed some high threshold:

$$\mathcal{N}_\mathbb{S} = \max_{\substack{s,t \in V \\ s \neq t}} \frac{\text{len}_G(\mathbb{S}(s,t))}{d_G(s,t)} \tag{4.9}$$

where $\mathbb{S}$ is the versatile routing scheme used for data routing through the network and will be described in the next section. Value $\mathcal{N}_{\mathbb{S}}$, which is called the network cost factor, presents an upper bound on the number of hops of a data flow routed by scheme $\mathbb{S}$ (which is equal to $\text{len}_G(\mathbb{S}(s, t))$) in comparison with the minimum possible number of hops ($d_G(s, t)$).

In the rest of this chapter, the versatile routing scheme of our model will be addressed in more detail. The two concerns brought up above will also be considered within our proposed routing scheme.

4.5   A Versatile Routing Scheme

As mentioned in section 4.2, the cyber devices in our hybrid model communicate with each other using four types of messages during different phases of the proposed protocol: REQ, NOTIF, ACK, and DATA. The first three types are *control* messages while the last one contains the requested data. There are relatively few control messages in a connection session; however, we may have multiple of DATA messages in a single connection. Additionally, the DATA messages are usually much larger in size than the control messages. Henceforth, we will restrict our routing cost discussion to DATA messages only.

As mentioned before, in order to route a DATA message in the data transmission phase, we need to first choose an intermediate routing node. Then, we use the location-based routing scheme to send the DATA message from the supplier to the intermediate node and subsequently, forward it toward the interested node which is the message's ultimate target. In this section, we describe the routing process in more detail. More specifically, we explain the way of choosing the intermediate node.

Busch's Randomized Algorithm   Here, we present an algorithm to compute a versatile routing scheme for our model. This algorithm, which was originally proposed by Busch et al. (2005) uses a randomized solution to make the traffic pattern of an existing routing scheme distributed (node congestion free) and meanwhile avoids a considerable increase in the total cost of data routing at the network level.

Remember that in our model, the network of routing nodes is represented by geometric graph $G = (V, E, \text{loc})$ which is deployed in convex area $\mathcal{A}$ belonging to the Euclidean plane. In addition, let symbol $\mathbb{Q}$ denote the existing location-based routing scheme, which is assumed to be implemented in the form of some non-centralized forwarding tables in the routing nodes. We call this scheme the default routing scheme. Algorithm 4.2

Algorithm 4.2 Randomized routing algorithm

Input: Vertices s and t (that are source and target vertices respectively)
Output: Randomized path $\mathbb{S}(s, t)$ between s and t in graph G

1    $\overline{ST} \leftarrow$ the line segment connecting points $S = \text{loc}(s)$ and $T = \text{loc}(t)$;
2    $\overline{UV} \leftarrow$ the perpendicular bisector line segment of $\overline{ST}$; /* Line segments a and b are perpendicular bisectors of each other if $a \perp b$ and each one bisects the other*/
3    $\overline{XY} \leftarrow \overline{UV} \cap \mathcal{A}$;
4    $x \leftarrow \text{IntermediateVertexSelection}(\overline{XY})$;
5    $\mathbb{S}(s, t) \leftarrow \mathbb{Q}(s, x) \oplus \mathbb{Q}(x, t)$;

---

gives fuller detail of the data routing process. Note that for every pair of vertices s, $t \in V$, we call path $\mathbb{Q}(s, t)$ the *default path* between s and t; in fact, we call the paths proposed by the existing routing scheme the default paths.

As shown in algorithm 4.2, $\overline{ST}$ denotes the line segment connecting the source and target vertices. Moreover, $\overline{UV}$ represents the perpendicular bisecting line segment of $\overline{ST}$, i.e., $\overline{UV}$ is perpendicular to $\overline{ST}$ and each one bisects the other (see figure 4.5). Since vertices s and t are located in a convex area $(\mathcal{A})$, the result of expression $\overline{UV} \cap \mathcal{A}$ is a line segment[5] $(\overline{XY})$. Moreover, function *Intermediate Vertex Selection* determines the intermediate vertex x. Note that the output path is obtained by merging paths $\mathbb{Q}(s, x)$ and $\mathbb{Q}(x, t)$ and is denoted by $\mathbb{S}(s, t)$. In other words, we use algorithm 4.2 to create a randomized routing scheme $\mathbb{S}$ for the given default routing scheme $\mathbb{Q}$. We address function *Intermediate Vertex Selection* in greater detail in algorithm 4.3.

In algorithm 4.3, point M is chosen randomly on the input line segment. Since the input line segment completely belongs to area $\mathcal{A}$, point M is also inside this area. Consequently, as area $\mathcal{A}$ is thoroughly $\mathcal{R}$-covered by the network of routing nodes, there exists a routing node at most $\mathcal{R}$ units away from point M. Lines 2–6 of the algorithm outline the search for the representing vertex of such a routing node.

In the next two sections, we analyze the described routing scheme in the context of the vertex-level cost (node congestion) and the network-level routing cost, which is evaluated using factor $\mathcal{N}_s$ defined by equation (4.9).

---

5. In fact, the value of $\overline{UV} \cap \mathcal{A}$ can also be a point; but we consider a point as a line segment of length zero.

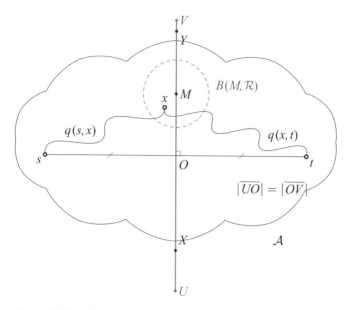

Figure 4.5 How the randomized algorithm finds a path between source s and target t.

Algorithm 4.3   Intermediate vertex selection

Input: Line segment l
Output: The intermediate vertex x inside the area $\mathcal{A}$
1    $M \leftarrow$ A uniformly distributed random point on l;
2   **foreach** vertex x of graph G
3   **if** $\mathrm{loc}(x) \in B(M, \mathcal{R})$ **then**
4   **return** x;
5   **end**
6   **end**   // x is one of the graph vertices that is in ball B(M, R)

4.6   Node Congestion Prevention

We now analyze the vertex-level cost of the randomized scheme presented in the previous section. To do this, we obtain an upper bound for the expected competitive ratio of the versatile routing scheme in the oblivious routing problem defined in section 4.4:

$$E[CR(\mathbb{E}, \mathbb{S})] = \frac{E[C_\mathbb{S}]}{C^*}$$

In the above equation, $\mathbb{E}$ is the set of possible routing cost environments defined by equation (4.8), C^* is the optimal cost of the oblivious routing problem, and C_S is the cost of the solution proposed by scheme S.

Before starting our discussion, some preliminary definitions will be presented.

4.6.1 Preliminary Definitions

For every convex subset in the Euclidean plane, there exists a number defined as its *pseudo-convexity factor*, which is formally defined in Definition 4.1.

Definition 4.1 If S denotes a convex set of points on the Euclidean plane, the *pseudo-convexity factor* γ^S is defined as:

$$\gamma^S = \inf_{A,B \in S} \frac{\left|\overline{AB}^{\perp} \cap S\right|}{\left|\overline{AB}\right|}$$

such that line segment $\overline{AB}^{\perp}$ denotes the perpendicular bisecting line segment[6] of $\overline{AB}$; and $\left|\overline{AB}^{\perp} \cap S\right|$ is the length of the part of line segment $\overline{AB}^{\perp}$ that is within area S.

In the above definition, note that as A and B belong to the convex set S, the result of expression $\overline{AB}^{\perp} \cap S$ is always a line segment or a point (line segment of length zero). The pseudo-convexity factor is shown for some familiar convex subsets of the Euclidean plane in figure 4.6. In this section,

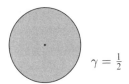

$\gamma = \frac{1}{2}$

(a) A circle of unit radius.

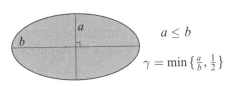

$a \leq b$

$\gamma = \min\left\{\frac{a}{b}, \frac{1}{2}\right\}$

(b) An ellipse of diameters a and b.

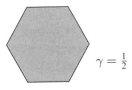

$\gamma = \frac{1}{2}$

(c) A regular hexagon.

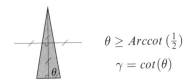

$\theta \geq Arccot\left(\frac{1}{2}\right)$

$\gamma = cot(\theta)$

(d) A bilateral triangle.

Figure 4.6 Pseudo-convexity of some familiar convex subsets of the Euclidean plane.

6. Line segment $\overline{AB}^{\perp}$ is the perpendicular bisecting line segment of $\overline{AB}$, if and only if $\overline{AB}^{\perp} \perp \overline{AB}$, $\overline{AB}$ bisects $\overline{AB}^{\perp}$, and $\overline{AB}^{\perp}$ bisects $\overline{AB}$.

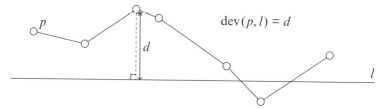

Figure 4.7 Deviation of path p from line l.

assume that γ represents the pseudo-convexity factor of convex area $\mathcal{A}$ over which the routing nodes are distributed.

Definition 4.2   Let $G = (V, E, \text{loc})$ denote a geometric graph in the Euclidean plane and p represent a path in G. *Deviation of path p from line l* in the plane is denoted by dev(p, l) and defined in the following form:

$$\text{dev}(p, l) = \max_{v \in N_p} \text{distance}(\text{loc}(v), l) \qquad (4.10)$$

such that set $N_p \subseteq V$ denotes the set of vertices participating in path p and distance (O, l) specifies the Euclidean distance[7] from point O to line l (see figure 4.7).

In the remainder of this section, we consider the following value (δ) to be the *deviation factor* of the default scheme $\mathbb{Q}$:

$$\delta = \max_{u,v \in V} dev\left(\mathbb{Q}(u, v), \overline{UV}\right)$$

such that $U = \text{loc}(u)$, $V = \text{loc}(v)$, and $\overline{UV}$ denotes the line passing through U and V. In fact, deviation factor δ specifies the maximum deviation of every default path from the straight line connecting its end vertices.

4.6.2   The Expected Competitive Ratio

At first, we compute a high threshold for the probability of using some vertex $v \in V$ in path $\mathbb{S}(s, t)$ between vertices s and t where $\mathbb{S}$ is the versatile scheme described in section 4.4. As mentioned before, path $\mathbb{S}(s, t)$ is the union of two default paths $\mathbb{Q}(s, x)$ and $\mathbb{Q}(x, t)$ such that x is the intermediate vertex selected by algorithm 4.3. If path $\mathbb{S}(s, t)$ passes through v, vertex v is either on the first part ($\mathbb{Q}(s, x)$) or on the second part ($\mathbb{Q}(x, t)$). We will focus on finding a high threshold for the probability that v is on $\mathbb{Q}(s, x)$; then, we extend our result to path $\mathbb{Q}(x, t)$.

Theorem 4.1   Let x denote the intermediate vertex of path $\mathbb{S}(s, t)$ in algorithm 4.2. The probability that vertex v participates in default path $\mathbb{Q}(s, x)$ has the following upper bound:

---

7. Euclidean distance from point O to line l is defined as the length of the line segment perpendicular to line l from point O.

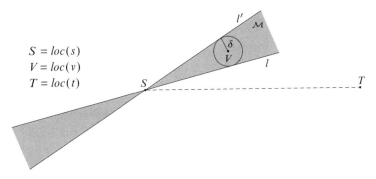

$$S = loc(s)$$
$$V = loc(v)$$
$$T = loc(t)$$

Figure 4.8 The highlighted area shows the set of points that vertex x can possibly be located in.

$$\mathbf{Pr}[v \text{ is on } \mathbb{Q}(s, x)] \leq \frac{5}{\gamma}\left(\frac{\mathcal{R}}{d_{\text{loc}}(s, t)} + \frac{\delta}{d_{\text{loc}}(s, v)}\right) \qquad (4.11)$$

Proof. Let $X, S \in \mathcal{A}$ denote the locations of vertices x and s, respectively. Equation (4.10) implies that:

$$\text{dev}\left(\mathbb{Q}(s, x), \overline{SX}\right) \leq \delta$$

Or equivalently,

$$\text{distance}\left(\text{loc}(z), \overline{SX}\right) \leq \delta$$

for every vertex z on path $\mathbb{Q}(s, x)$. The last inequality implies that if path $\mathbb{Q}(s, x)$ passes through vertex v, the Euclidean distance of point $V = \text{loc}(v)$ from line $\overline{SX}$ will be less than or equal to δ. As a result, point X must belong to the area $\mathcal{M}$ highlighted in figure 4.8 (note that in this figure, lines l and l' pass through S and are tangent to the circle of radius δ and center V).

Moreover, concerning algorithm 4.3, vertex x must also be in ball $B(M, \mathcal{R})$ where M is a randomly distributed point on $\overline{S'T'}$, which is the perpendicular bisecting line segment of ST. As a result, point X should also be located in area $\mathcal{M}'$, which implies that:

$$\left.\begin{array}{c} X \in \mathcal{M} \\ X \in \mathcal{M}' \end{array}\right\} \rightarrow X \in \mathcal{M} \cap \mathcal{M}'$$

Area $\mathcal{M} \cap \mathcal{M}'$ has been shaded in figure 4.9. Additionally, as the distance $\|X - M\|$ is not larger than $\mathcal{R}$, point M can only be located on line segment $\overline{AB}$ specified in figure 4.9. Consequently, we obtain the following proposition:

$$(v \text{ is on } \mathbb{Q}(s, x)) \rightarrow \left(M \in \overline{AB}\right)$$

which implies that:

$$\mathbf{Pr}[v \text{ is on } \mathbb{Q}(s, x)] \leq \mathbf{Pr}\left[M \in \overline{AB}\right] \qquad (4.12)$$

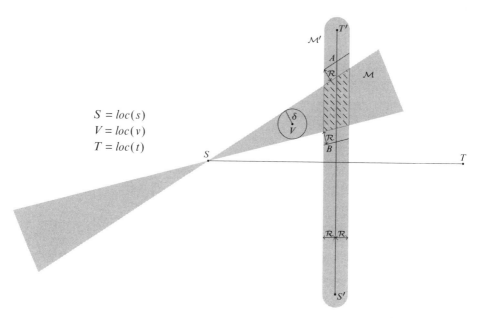

$$S = loc(s)$$
$$V = loc(v)$$
$$T = loc(t)$$

Figure 4.9 The gray line segment specifies the set of points to which point M belongs.

As point M is uniformly distributed over line segment $\overline{S'T'} \cap \mathcal{A}$, we conclude that:

$$\mathbf{Pr}\big[M \in \overline{AB} \big] = \frac{\overline{AB}}{\big| \overline{S'T'} \cap \mathcal{A} \big|} = \frac{\overline{AB}}{\big| \overline{S'T'} \big| \times \dfrac{\big| \overline{S'T'} \cap \mathcal{A} \big|}{\big| \overline{S'T'} \big|}}$$

Moreover, regarding the definition of the pseudo-convexity factor, we obtain the following inequality:

$$\gamma \le \frac{\big| \overline{S'T'} \cap \mathcal{A} \big|}{\big| \overline{ST} \big|} \le \frac{\big| \overline{S'T'} \cap \mathcal{A} \big|}{\big| \overline{S'T'} \big|}$$

Henceforth, we obtain the following relation:

$$\mathbf{Pr}\big[M \in \overline{AB} \big] \le \frac{\big| \overline{AB} \big|}{\gamma \big| \overline{S'T'} \big|} \le \frac{\big| \overline{AB} \big|}{\gamma d_{\text{loc}}(s, t)}$$

Regarding inequality (4.12), we conclude that:

$$\mathbf{Pr}[v \text{ is on } \mathbb{Q}(s, x)] \le \frac{\big| \overline{AB} \big|}{\gamma \times d_{\text{loc}}(s, t)}$$

In the exercises, you will be asked to geometrically prove the following upper bound for the value of $|\overline{AB}|$:

$$|\overline{AB}| < 5\left(\mathcal{R} + \frac{\delta|\overline{ST}|}{|\overline{SV}|}\right)$$

(4.13)

As a result, we obtain inequality (4.11). ∎

Concerning Theorem 4.1, the expected value of $\mathrm{CR}(\mathbb{E}, \mathbb{S})$ is bounded to the following value:

$$\mathrm{E}[\mathrm{CR}(\mathbb{E}, \mathbb{S})] = \Theta\left(c\mathcal{R} \cdot \Psi(G) \cdot \max\{\mathcal{R}, \delta\}^2\right)$$

(4.14)

In the exercises, you will be asked to prove equation (4.14).

4.7 Routing Cost Analysis

In this section, we address the network-level cost issues of routing scheme $\mathbb{S}$ presented in section 4.5. In fact, we obtain an upper bound for the value of the network cost factor $\mathcal{N}_\mathbb{S}$ defined in equation (4.9). This upper bound is a scaled value of the network cost factor of the default routing scheme ($\mathcal{N}_\mathbb{Q}$).

Let x denote the intermediate vertex chosen by algorithm 4.2 when the input vertices are s and t. Since the randomized output path $\mathbb{S}(s, t)$ is equal to $\mathbb{Q}(s, x) \cup \mathbb{Q}(x, t)$, we deduce that:

$$\mathrm{len}_G(\mathbb{S}(s, t)) = \mathrm{len}_G(\mathbb{Q}(s, x)) + \mathrm{len}_G(\mathbb{Q}(x, t))$$

Moreover, given that $\mathcal{N}_\mathbb{Q}$ is the network-level cost factor of default scheme $\mathbb{Q}$, we obtain the following inequalities:

$$\frac{\mathrm{len}_G(\mathbb{Q}(s, x))}{d_G(s, x)} \leq \mathcal{N}_\mathbb{Q}$$

$$\frac{\mathrm{len}_G(\mathbb{Q}(x, t))}{d_G(x, t)} \leq \mathcal{N}_\mathbb{Q}$$

The above relations lead us to the following inequality for $\mathrm{len}_G(\mathbb{S}(s, t))$.

$$\mathrm{len}_G(\mathbb{S}(s, t)) \leq \mathcal{N}_\mathbb{Q}(d_G(s, x) + d_G(x, t))$$

(4.15)

In the next step, we will find an upper bound for the RHS[8] expression of inequality (4.15). Given that $\Psi(G)$ is the pseudo-diameter of G, the following inequalities hold:

$$\frac{d_G(s, x)}{d_{\mathrm{loc}}(s, x)} \leq \Psi(G) \to d_G(s, x) \leq \Psi(G) \cdot d_{\mathrm{loc}}(s, x)$$

---

8. Right-hand side.

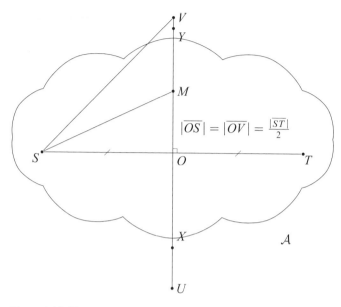

Figure 4.10 The geometric approach for computing an upper bound for the network-level cost factor $\mathcal{N}_s$.

On the other hand, concerning figure 4.10, if M denotes the point chosen in line 1 of algorithm 4.3, and $S = \text{loc}(s)$, we deduce that:

$$d_{\text{loc}}(s, x) = \|S - \text{loc}(x)\|$$

and

$$\|\text{loc}(x) - M\| \leq \mathcal{R}$$

As a result, regarding the triangular inequality, we obtain the following relation:

$$d_{\text{loc}}(s, x) \leq |\overline{SM}| + \|\text{loc}(x) - M\| \leq |\overline{SM}| + \mathcal{R}$$

In addition, regarding figure 4.10 and the Pythagorean theorem, it can be inferred that:

$$|\overline{SM}|^2 = |\overline{OM}|^2 + |\overline{SO}|^2$$

Without loss of generality, we assume that M is a point on line segment $\overline{YO}$ (or similarly, it may be on $\overline{XO}$). Since $\overline{XY} \subseteq \overline{UV}$, we obtain that $|\overline{OM}| \leq |\overline{OY}| \leq |\overline{OV}|$; consequently, we obtain the following inequality:

$$|\overline{SM}|^2 \leq |\overline{OM}|^2 + |\overline{SO}|^2 \leq 2\left(\frac{|\overline{ST}|}{2}\right)^2$$

In the above relations, note that $|\overline{OV}| = |\overline{SO}| = \dfrac{|ST|}{2}$.

As a result, we obtain the following inequality regarding $d_{\text{loc}}(s, x)$:

$$d_{\text{loc}}(s, x) \le \frac{1}{\sqrt{2}}|\overline{ST}| + \mathcal{R}$$

which leads to the following upper bound on the value of $d_G(s, x)$:

$$d_G(s, x) \le \Psi(G)\frac{\sqrt{2}|\overline{ST}| + 2\mathcal{R}}{2}$$

In a similar way, we conclude the following high threshold for $d_G(x, t)$:

$$d_G(x, t) \le \Psi(G)\frac{\sqrt{2}|\overline{ST}| + 2\mathcal{R}}{2}$$

As a result, according to inequality (4.15), the following inequalities hold:

$$\text{len}_{G(\mathbb{S}(s,t))} \le \mathcal{N}_{\mathbb{Q}}\Psi(G)\left(\sqrt{2}|\overline{ST}| + 2\mathcal{R}\right) \le \mathcal{N}_{\mathbb{Q}}\Psi(G)\left(\sqrt{2}d_{\text{loc}}(s, t) + 2\mathcal{R}\right)$$

At the next step, we will find a relation between $d_{\text{loc}}(s, t)$ and $d_G(s, t)$. Considering p as the shortest path from s to t in G, the value $d_G(s, t)$ is by definition equal to $|p|$. Moreover, using the triangular inequality, we obtain the following relation:

$$d_{\text{loc}}(s, t) \le \delta_G(s, t) \le \sum_{\{u,v\}\in p} d_{\text{loc}}(u, v) \le \sum_{\{u,v\}\in p} c\mathcal{R} = c\mathcal{R}\sum_{\{u,v\}\in p} 1 = c\mathcal{R}d_G(s, t)$$

Consequently, we find an upper bound for $\text{len}_G(\mathbb{S}(s,t))$ in the following form:

$$\text{len}_G(\mathbb{S}(s, t)) \le \mathcal{N}_{\mathbb{Q}}\Psi(G)\left(\sqrt{2}c\mathcal{R}d_G(s, t) + 2\mathcal{R}\right)$$

$$= \mathcal{N}_{\mathbb{Q}}\Psi(G)\mathcal{R}\left(\sqrt{2}c + \frac{2}{d_G(s, t)}\right)d_G(s, t)$$

Since $d_G(s, t)$ is a positive integer, $d_G(s, t) \ge 1$; henceforth:

$$\text{len}_G(\mathbb{S}(s, t)) \le \mathcal{N}_{\mathbb{Q}}\Psi(G)\mathcal{R}\left(\sqrt{2}c + 2\right)d_G(s, t)$$

Finally, we obtain the following upper bound for $\mathcal{N}_{\mathbb{S}}$:

$$\mathcal{N}_{\mathbb{S}} = \frac{\text{len}_G(\mathbb{S}(s, t))}{d_G(s, t)} \le \left(2 + \sqrt{2}c\right)\mathcal{R}\Psi(G)\mathcal{N}_{\mathbb{Q}} = \Theta(R\Psi(G))\mathcal{N}_{\mathbb{Q}}$$

or equivalently,

$$\frac{\mathcal{N}_{\mathbb{S}}}{\mathcal{N}_{\mathbb{Q}}} \le \Theta(R\Psi(G)) \tag{4.16}$$

As seen in inequality (4.16), the network cost factor of the versatile scheme obtained by algorithm 4.2 will not be $\Theta(R\Psi(G))$ times more than the cost factor of the default scheme.

4.8 Summary and Outlook

In this chapter, we developed a hybrid model of a peer-to-peer network that provides a secure congestion-free way of distributing data over a network of routing nodes deployed in a convex subset of the Euclidean plane. This model uses a content-centric protocol for sending the requests; while a host-based scheme is used for routing the data flow through the network.

We used a security scheme for the model to manage the security issues based on the content being transformed. Additionally, the routing scheme applied in the model makes the data traffic pattern of our network distributed and node congestion free.

Exercises

1. Considering figure 4.9, use geometrical deduction to show inequality (4.13).
2. Prove equation (4.14) (*Hint*: Use the approach proposed in Busch, Madgon-Ismail, and Xi, 2005).
3. Show that for every bounded subset of the Euclidean plane, the pseudo-convexity factor is *not* more than 1/2.

Suggested Reading

Busch, C., M. Magdon-Ismail, and J. Xi. 2005. Oblivious routing on geometric networks. *Proceedings of the 17th ACM Symposium on Parallelism in Algorithms and Architectures, SPAA*, 316–324.

Carter, K. M., R. P. Lippmann, and S. W. Boyer. 2010. Temporally oblivious anomaly detection on large networks using functional peers. *Proceedings of the 10th ACM Sig-Comm Conference on Internet Measurement*, 465–471.

Chan, H., A. Perrig, and D. Song. 2003. Random key predistribution schemes for sensor networks. *Proceedings of the IEEE Symposium on Security and Privacy*, 197.

Jeong, J., T. T. Kwon, and Y. Choi. 2010. Host-oblivious security for content-based networks. *Proceedings of the 5th International Conference on Future Internet Technologies*, 35–40.

Kurosawa, K., W. Kishimoto, and T. Koshiba. 2008. A combinatorial approach to deriving lower bounds for perfectly secure oblivious transfer reductions. *IEEE Transactions on Information Theory*, 54(6):2566–2571.

Chapter 5   Versatile Distribution of Green Power Resources

There has been a growing trend in the electric power distribution from a centralized producer-driven grid to a smarter interactive consumer. This requirement compels a new way of designing smart grids for efficient power distribution. Recent advances in wireless communication and smart grid design have enabled the development of low cost power distribution systems. In a related context, the negative effects of coal plants will create a major environmental hazard in addition to changing the design of power plants. Henceforth, not only is the distribution of electric power approach changing but the producers of the electricity are also focusing on cleaner methods for guaranteeing a lower bound throughput.

One of the most critical challenges of designing smarter grids is to make the power distribution facilities more flexible to the variable nature of the electricity consuming and generating patterns. More specifically, producing power using a green plant such as solar does not guarantee a low threshold of energy throughput within a certain amount of time. In addition, consumers have a time-varying usage pattern in the context of an electricity residential system. By considering all of these uncertainties, designing a versatile power distribution scheme is absolutely fundamental to develop a reliable grid. In other words, the goal is to create a power distribution grid which can satisfy the consumers (with high probability) by providing enough electricity for them at any given moment of time; however, the amount of energy production of green power plants may vary widely from time to time.

In this chapter, we introduce a novel and robust distribution scheme to transfer the electric energy generated by power plants to residential consumers through a smart grid system. Additionally, the distribution method will be analyzed in the context of reliability and cost efficiency. To accomplish this paradigm, we provide an optimal stochastic framework to mathematically formulate and model the problem. Then, we use a versatile routing scheme to fairly balance the electricity load through the smart grid. Additionally, we use a computational algorithm for the global minimization of a linearly constrained concave function by partition of the feasible domain.

5.1 Introduction

As an alternative to centralized power production, distributed generation is the use of small-scale power generation technologies located near the load being served, capable of lowering costs, improving reliability, reducing emissions, and expanding energy options. Such distribution facilities are characterized by a two-way flow of electricity and information and are also capable of monitoring everything from power plants, to customer preferences, to individual appliances. Additionally, the dream of instantaneous balance of supply and demand (at the device level) in the electric industry is being turned into reality by deploying these smart grids, as shown in figure 5.1.

We now present an example of smart grids in the context of a residential electricity distribution grid.

A Residential Electricity System Consider a network of communities in a city. In each community, there are a number of electricity consumers and a *renewable power plant*[1] that supplies the energy needed by the electricity consumers in the community. Moreover, there are a few power plants outside the communities and scattered over the city to help the renewable power plants generate electricity on demand. The existence of these extra plants improves the reliability of our electrical distribution system. We refer to these extra power plants as *auxiliary power plants*. The electric energy generated by these generators can be transferred to each community in the city through the network of communities schematically represented in figure 5.2. This figure shows a network representation of our described model. Note that each community has some connected neighbors so that it can trade electricity with them if necessary. We assume that any two neighbors are connected via a two-way transmission power line.

Consider the community in our example. Each electricity consumer located in the community has some amount of electricity demand, which varies from time to time. Let $\mathrm{Dem}(t)$ denote the total electricity demand of all of the consumers in the community at given time t. Note that $\mathrm{Dem}(t)$ is

---

1. A renewable power plant is an industrial facility for the generation of electric power. The power generator of a renewable power plant uses renewable energy sources.

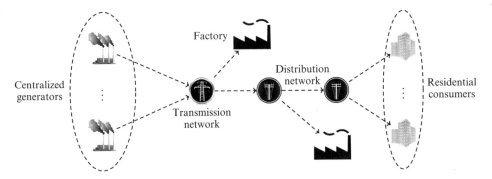

(a) Centralized, producer-controlled grid of power distribution in the past.

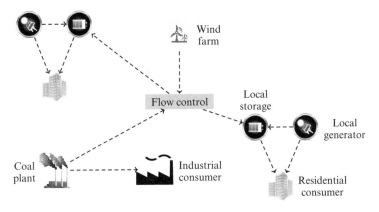

(b) Clean, local, and consumer-interactive grid of power distribution in the future.

Figure 5.1 The transformation happening in the electrical power distribution grids.

the instantaneous power that has to be supplied at time t. Additionally, we use Gen(t) to represent the total amount of electric power generated by the power plant located in the community at a given time t.

Figure 5.3 shows the changes of residential power demand Dem(t) on a specific day in New England.

Figure 5.4 shows more detail about the community. Note that there is a controller unit that controls the energy flow inside the community. Furthermore, the controller may participate in the energy distribution of other communities by forwarding the flow received from its neighbors. Assuming that the production amount Gen(t) of the power plant exceeds Dem(t) in a period of time, the energy flow of size Gen(t) from the power

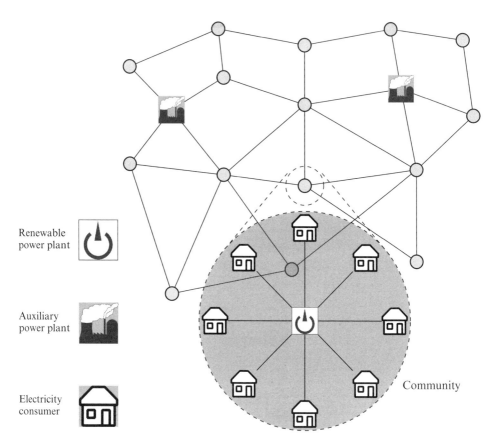

Renewable
power plant

Auxiliary
power plant

Community

Electricity
consumer

Figure 5.2 Schematic representation of the electric power distribution network.

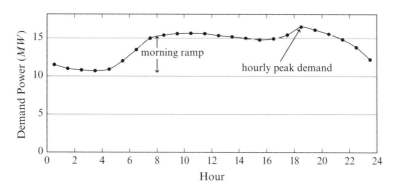

Figure 5.3 Residential power demand in New England, 10/22/2010.

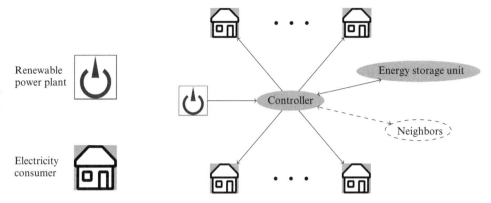

Figure 5.4 Schematic view of a community. Arrows represent energy flow.

plant is divided into two parts: a power flow of size Dem(t) moves toward the consumers and the power flow Gen(t) − Dem(t) is stored by the energy storage unit embedded in each community. If the value of Gen(t) becomes lower than Dem(t) for some time interval, there will be two main energy flows in the community to satisfy consumer demand: one is from the power plant of size Gen(t) and the other is from the storage unit. Moreover, in the case that Dem(t) > Gen(t) and where there is not enough energy in the storage unit to compensate for the shortage of power plant production, there has to be another flow from the neighbors toward the community consumers. Later in this section, we will address each case in more detail.

5.2 Electricity System Reliability

In a smart grid system, the power needed by the consumers of a community is supplied by the generators and the storage unit embedded in the community. A system that provides energy is reliable if it satisfies consumer demand most of the time. More specifically, in our example, the system is reliable if in the intervals that the instantaneous power demand Dem(t) exceeds the power supplied by the community power plant at time t, there will exist enough stored energy in the community to cover the exceeded energy demand.

In this section, we discuss reliability issues of our electric system using a reasonable stochastic model of the amount of supplied and demanded energy. Before starting the discussion, we briefly review some concepts regarding our stochastic processes.

5.2.1 Stochastic Processes

Stochastic process X_t is a function of domain $D \subseteq \mathbb{R}_{\geq 0}$ such that for every $t \in D$, X_t is a random variable. Stochastic processes are used to model the evolution of a random variable over time. The deterministic function $\mu : D \mapsto \mathbb{R}$ is the *mean function* of process X_t if:

$$\mu_t = \mathbf{E}[X_t] \quad \forall t \in D$$

Additionally, the *covariance function* of process X_t is defined by the following equation:

$$\mathbf{cov}(s, t) = \mathbf{E}[(X_s - \mu_s)(X_t - \mu_t)]$$

Process B_t is the *Brownian Motion Process* if the following three conditions hold:

1. The map $t \mapsto B_t(\omega)$ is continuous for every ω.
2. Assuming that $t_1 \leq t_2 \leq \dots \leq t_n$, the random variables belonging to the following set are mutually independent:

$$\{(B_{t_{i+1}} - B_{t_i}) : i = 1, 2 \dots, n-1\}$$

3. Every process increment is stationary, i.e., the probability distribution of $B_t - B_s$ only depends on $t - s$ for every $s, t > 0$.

For every $t > 0$, the random variable $B_t - B_0$ can be written as n i.i.d.[2] random variables in the following form:

$$B_t - B_0 = \sum_{i=1}^{n} B_{\frac{it}{n}} - B_{\frac{(i-1)t}{n}}$$

Concerning the central limit theorem, the above summation converges to a normally distributed random variable if $n \to +\infty$. Moreover, we can prove that $(B_t - B_0) \sim \mathcal{N}(\mu t, \sigma^2 t)$ for some values of μ and σ^2. In the case that $\mu = 0$, $\sigma^2 = 1$, and $B_0 = 0$, process B_t is known as the *Wiener process*. The Wiener process of $\sigma^2 \neq 1$ is called the *scaled* Wiener process of variance σ^2. We will address some properties of the scaled Wiener process later.

If W_t denotes a scaled Wiener process of variance σ^2, $\mu_t = 0$ for every $t \geq 0$, and

$$\mathbf{cov}(s, t) = \sigma^2 \min\{s, t\} \quad \forall s, t \geq 0$$

Furthermore, process M_t is known as the *running maximum* of Wiener process W_t, if:

$$M_t = \sup_{s \in [0,t]} W_s$$

2. Independent, identically distributed.

For such a process, the cumulative density function is obtained by the following equation:

$$\Pr[M_t \leq m] = \mathrm{erf}\left(\frac{m}{\sqrt{2t\sigma^2}}\right)$$

Additionally,

$$\mathrm{E}[M_t] = \sqrt{\frac{2t\sigma^2}{\pi}}$$

and

$$\mathrm{var}(M_t) = \left(1 - \frac{2}{\pi}\right)t\sigma^2$$

The generalized derivative $w_t = \dot{W}_t$ of the scaled Wiener process W_t is known as the *Gaussian white noise* process. In other words,

$$\int_0^\infty g(t)w_t\mathrm{d}t = -\int_0^\infty \dot{g}(t)w_t\mathrm{d}t \quad \text{for every smooth function } g$$

For process w_t as a Gaussian white noise process, the mean and covariance functions are obtained by the following equations:

$$\mu_t = 0 \quad \forall t \geq 0$$

$$\mathrm{cov}(s,t) = \sigma^2\delta(s-t) \quad \forall s, t \geq 0$$

5.2.2 Modeling with Stochastic Processes

Here, we develop a reasonable stochastic model for the amount of energy consumption and production of a single community. At first, we assume that functions $\mathrm{Dem}(t)$ and $\mathrm{Gen}(t)$ are approximately periodic with period T, i.e.:

$$\mathrm{Dem}(t+T) \simeq \mathrm{Dem}(t) \wedge \mathrm{Gen}(t+T) \simeq \mathrm{Gen}(t) \quad \forall t \in [0, T)$$

In addition, assume that we have already analyzed the instantaneous demand and supply of each community over time interval $[0, nT)$ for some large n.

Now, consider the following average functions which are reasonable estimations of functions $\mathrm{Dem}(t)$ and $\mathrm{Gen}(t)$ (based on historical data):

$$\begin{cases} \mathrm{Dem}_\mu(t) = \lim_{n \to +\infty} \sum_{i=0}^{n-1} \dfrac{\mathrm{Dem}(t+iT)}{n} \\ \mathrm{Gen}_\mu(t) = \lim_{n \to +\infty} \sum_{i=0}^{n-1} \dfrac{\mathrm{Gen}(t+iT)}{n} \end{cases} \quad \forall t \in [0, T)$$

We call $\mathrm{Dem}_\mu(t)$ and $\mathrm{Gen}_\mu(t)$ the expected values of power demand and supply, respectively, in the community at time t. Note that functions $\text{Dem}_\mu(t)$ and $\text{Gen}_\mu(t)$ are periodically defined for $t \geq T$, i.e.:

$$\text{Dem}_\mu(t) = \text{Dem}_\mu(t-T) \wedge \text{Gen}_\mu(t) = \text{Gen}_\mu(t-T) \quad \forall t \geq T$$

Now, we model functions $\text{Dem}(t)$ and $\text{Gen}(t)$ as two random processes in the following form:

$$\begin{cases} \text{Dem}(t) = \text{Dem}_\mu(t) + D_t \\ \text{Gen}(t) = \text{Gen}_\mu(t) + G_t \end{cases} \quad \forall t \geq 0 \tag{5.1}$$

where D_t and G_t are two independent Gaussian white noises of the following covariance functions.

$$\begin{cases} \mathbf{cov}_D(s,t) = \sigma_d^2 \cdot \delta(s-t) \\ \mathbf{cov}_G(s,t) = \sigma_g^2 \cdot \delta(s-t) \end{cases} \quad \forall s, t \geq 0$$

At a given time t, each community has some amount of stored energy represented by $\mathcal{S}(t)$, which is obtained by the following equation:

$$\mathcal{S}(t) = \int_0^t (\text{Gen}(t') - \text{Dem}(t')) dt' + s_0 \quad \forall t \geq 0 \tag{5.2}$$

such that s_0 is the initial stored energy in the community. Note that in this section, we assume that each community does not get electric energy from the auxiliary power plants located outside of the community. By replacing the supply and demand functions with their equivalent random processes, we obtain that:

$$\begin{aligned} \mathcal{S}(t) &= \int_0^t (\text{Gen}_\mu(t') - \text{Dem}_\mu(t') + G_{t'} - D_{t'}) dt' + s_0 \\ &= \int_0^t (\text{Gen}_\mu(t') - \text{Dem}_\mu(t')) dt' + s_0 - \int_0^t (D_{t'} - G_{t'}) dt' \\ &= \mathcal{S}_\mu(t) - W_t \end{aligned}$$

where:

$$\mathcal{S}_\mu(t) = \int_0^t (\text{Gen}_\mu(t') - \text{Dem}_\mu(t')) dt' + s_0$$

and

$$W_t = \int_0^t (D_{t'} - G_{t'}) dt'$$

Since G_t and D_t are two independent Gaussian white noises, W_t is a scaled Wiener process of variance $(\sigma_g^2 + \sigma_d^2)$. Henceforth, the covariance function of process W_t is in the following form:

$$\mathbf{Cov}(s,t) = \min\{s,t\} \cdot (\sigma_g^2 + \sigma_d^2)$$

Additionally, $\mathcal{S}_\mu(t)$ is the expected value of the stored energy at given time t:

$$\mathbf{E}[\mathcal{S}(t)] = \mathbf{E}[\mathcal{S}_\mu(t) - W_t] = \mathcal{S}_\mu(t) - \mathbf{E}[W_t] = \mathcal{S}_\mu(t) - 0 = \mathcal{S}_\mu(t)$$

According to the above analysis, the amount of stored energy $S(t)$ is equal to the summation of deterministic amount $S_\mu(t)$ and the scaled Wiener process $(-W_t)$. In the rest of this chapter, we assume that the expected value of the stored energy never becomes less than the initial amount of energy (s_0); i.e.:

$$S_\mu(t) \geq s_0 \quad \forall t \geq 0$$

5.2.3 Reliability Analysis

The probability that the actual stored energy $S(t')$ does not meet the low threshold $(s_0 - \delta)$ for some $\delta \in [0, s_0]$ and every $t' \leq t$ is:

$$\Pr[\forall t' \leq t : S(t') > s_0 - \delta] = \Pr[\forall t' \leq t : S_\mu(t') - W_{t'} > s_0 - \delta]$$
$$\geq \Pr[\forall t' \leq t : W_{t'} < \delta]$$
$$\geq \Pr[\sup_{t' \leq t} W_{t'} < \delta] = \Pr[M_t < \delta]$$

where M_t is the running maximum process corresponding to the scaled Wiener process W_t. As a result, we obtain the following inequality:

$$\Pr[\forall t' \leq t : S(t') > s_0 - \delta] \geq \text{erf}\left(\frac{\delta}{\sqrt{2t\sigma^2}}\right) \tag{5.3}$$

such that $\sigma^2 = \sigma_g^2 + \sigma_a^2$. We call $\Pr[\forall t' \leq t : S(t') > s_0 - \delta]$ as the *reliability ratio* of the electrical system of each community in time interval $[0,t]$. In other words, we assume that if $S(t) \leq s_0 - \delta$ (the stored energy becomes lower than some threshold), the consumer demand will no longer be satisfied. Figure 5.5 shows how the lower bound of the reliability ratio, obtained in inequality (5.3), changes as parameters δ and σ^2 get different values.

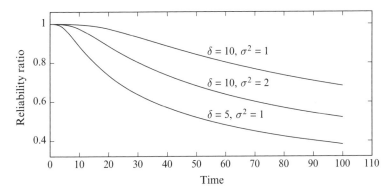

Figure 5.5 Lower bound of the reliability ratio of a community power plant for three different values of δ and σ^2.

As shown in figure 5.5, if the renewable power plant of each community is the only source of power for the community consumers, the reliability ratio will substantially decrease over time. Subsequently, we have to get help from the auxiliary power plants located outside the community to improve the reliability ratio by generating extra energy on demand.

Impact of Auxiliary Power Plants on System Reliability Consider the following decision making algorithm for each community which specifies when it has to ask the auxiliary power plants for extra energy and also determines how much energy has to be flowed into the community in that time. In algorithm 5.1, we assume that in time $t = T_1$, the amount of stored energy $S(t)$ becomes as low as $s_0 - \delta'$ units for the first time such that δ' is a positive number in interval $\left[\frac{\delta}{2}, \delta\right)$. Then, in period $[T_1, T_1 + d]$, the auxiliary power plants in the city flow $S_\mu(T_1) - s_0 + \delta'$ units of electric energy into the community. Note that for simplicity, we assume that the energy provided by the auxiliary power plants will be ready to use only after time $t = T_1 + d$. In other words, function $S(t)$ is not continuous in time $t = T_1 + d$:

$$\lim_{\varepsilon \to 0^+} S(T_1 + d + \varepsilon) - \lim_{\varepsilon \to 0^-} S(T_1 + d + \varepsilon) = S_\mu(T_1) - s_0 + \delta'$$

Algorithm 5.1   Community-level controller

/* Function Now() specifies how long it takes since the beginning of the algorithm. t_{end} is the latest moment we are concerned about*/

1    $i \leftarrow 1$;
2   **while** Now() $< t_{end}$ **do**
3   **if** $S(\text{Now}()) \neq s_0 - \delta'$ // δ' is some constant positive value in
               $\left[\frac{\delta}{2}, \delta\right]$.
4   **then**
5 | **continue**;
6   **end**
7          $T_i \leftarrow \text{Now}()$;
8   Get $(S_\mu(T_i) - s_0 + \delta')$ units of energy from external power plants
               in the interval $[T_i, T_i + d]$;
               // d is a deterministic positive number
9          $i \leftarrow i + 1$;
10   **end**

Again, we assume that T_2 is the smallest number greater than $T_1 + d$ such that $S(T_2) = s_0 - \delta'$. The variables $T_3, T_4, \ldots$ are also defined in the same way.

Note that the amount of energy flow in the ith iteration of the algorithm has the following deterministic upper bound:

$$S_\mu(T_i) - s_0 + \delta' \le S_{max} - s_0 + \delta'$$

where S_{max} denotes the maximum value of $S_\mu(t)$ for every $t \in [0, t_{end}]$ (the whole time interval that we are interested in). Subsequently, it is a feasible assumption that value d, which is the maximum duration of the energy flow in each iteration, is deterministic. In other words, we assume that the cables between communities have been designed in a way that the time needed to flow $(S_{max} - s_0 + \delta')$ units of energy into a community is at most d. Moreover, we will show in Theorem 5.1 that if the system remains reliable in time interval $[T_i, T_i + d]$, the actual stored energy at moment $t = T_i + d$ is definitely greater than $s_0 - \delta'$ for every possible i. This implies that the value of expression $S(\text{Now}(\ ))$ in line 3 of algorithm 5.1 never falls below $(s_0 - \delta')$; i.e.:

$$S(t) > s_0 - \delta' \quad \forall t \notin \bigcup_{\substack{i \ge 1 \\ T_i \le t_{end}}} [T_i, T_i + d] \tag{5.4}$$

Consequently, in the case that the If-statement in line 3 is wrong, the value of $S(\text{Now}(\ ))$ will be equal to $s_0 - \delta'$.

Since δ' is less than δ, according to inequality (5.4), the amount of stored energy $S(t)$ does not become less than $s_0 - \delta$ for every t satisfying inequality (5.4). Henceforth, in order to analyze the reliability of the electric system controlled by algorithm 5.1, we only need to find a lower bound for the following probability:

$$\Pr[\forall t \in [T_i, T_i + d] : S(t) > s_0 - \delta] \quad \forall i$$

Theorem 5.1 Assuming that algorithm 5.1 controls the electric system of communities, the following inequality holds:

$$\Pr[\forall t \in [T_i, T_i + d] : S(t) > s_0 - \delta] \ge erf\left(\frac{\delta - \delta' - d\gamma}{\sqrt{2d\sigma^2}}\right) \tag{5.5}$$

where $\gamma = \max_t(\text{Dem}_\mu(t) - \text{Gen}_\mu(t))$ and $Ti \le t_{end}$. Moreover, assuming that $S(T_i + d) > s_0 - \delta$, we obtain that:

$$\lim_{\varepsilon \to 0^+} S(T_i + d + \varepsilon) > s_0 - \delta'$$

Proof. Concerning equation (5.2) and algorithm 5.1, the value of $S(t)$ is equal to the following expression:

$$S(t) = s_0 + \int_0^t (\text{Gen}(t') - \text{Dem}(t')) dt' + S_{aux}(t)$$

such that $S_{\text{aux}}(t)$ denotes the total amount of energy provided by the auxiliary power plants in time interval $[0, t]$. Since $S_{\text{aux}}(t)$ is an increasing function, we obtain the following lower bound for $S(T_i + x)$ for every $x \in [0, d]$:

$$S(T_i + x) = s_0 + \int_0^{T_i+x} (\text{Gen}(t') - \text{Dem}(t'))\,dt' + S_{\text{aux}}(T_i + x)$$

$$\geq s_0 + \int_0^{T_i+x} (\text{Gen}(t') - \text{Dem}(t'))\,dt' + S_{\text{aux}}(T_i)$$

$$\geq S(T_i) + \int_{T_i}^{T_i+x} (\text{Gen}(t') - \text{Dem}(t'))\,dt'$$

$$\geq s_0 - \delta' + \int_{T_i}^{T_i+x} (\text{Gen}(t') - \text{Dem}(t'))\,dt'$$

$$\geq s_0 - \delta' + \int_{T_i}^{T_i+x} (\text{Gen}_\mu(t') - \text{Dem}_\mu(t'))\,dt' - \int_{T_i}^{T_i+x} (D_{t'} - G_{t'})\,dt'$$

In addition, as $\gamma = \max_t \text{Dem}_\mu(t) - \text{Gen}_\mu(t)$, we infer that:

$$\int_{T_i}^{T_i+x} (\text{Gen}_\mu(t') - \text{Dem}_\mu(t'))\,dt' \geq -x\gamma$$

Consequently, we obtain the following lower bound for the value of $S(T_i + x)$:

$$S(T_i + x) \geq s_0 - \delta' - x\gamma - \int_{T_i}^{T_i+x} (D_{t'} - G_{t'})\,dt' \tag{5.6}$$

Inequality (5.19) implies that:

$$\mathbf{Pr}[\forall x \in [0, d] : S(T_i + x) > s_0 - \delta]$$

$$\geq \mathbf{Pr}\left[s_0 - \delta' - x\gamma - \int_{T_i}^{T_i+x} (D_{t'} - G_{t'})\,dt' > s_0 - \delta \right]$$

$$\geq \mathbf{Pr}\left[\forall x \in [0, d] : \int_{T_i}^{T_i+x} (D_{t'} - G_{t'})\,dt' < \delta - \delta' - x\gamma \right]$$

Concerning the assumption that G_t and D_t are two independent Gaussian white noises, expression $\int_{T_i}^{T_i+x} (D_{t'} - G_{t'})\,dt'$ results in a scaled Wiener process of covariance $\min\{s, t\} \cdot \sigma^2$. This process is equal to the increment of random process W_t in interval $[T_i, T_i + x]$. Assume that:

$$W_x^{(i)} = \int_{T_i}^{T_i+x} (D_{t'} - G_{t'})\,dt'$$

and also $M_t^{(i)}$ denotes the corresponding running maximum process of $W_x^{(i)}$. As a result, we obtain the following inequalities:

$$\mathbf{Pr}[\forall x \in [0, d] : S(T_i + x) > s_0 - \delta] \geq \mathbf{Pr}[\forall x \in [0, d] : W_x^{(i)} < \delta - \delta' - x\gamma]$$

$$\geq \mathbf{Pr}\left[\sup_{x \in [0,d]} W_x^{(i)} < \delta - \delta' - d\gamma \right]$$

$$\geq \mathbf{Pr}[M_d^{(i)} < \delta - \delta' - d\gamma]$$

$$\geq \text{erf}\left(\frac{\delta - \delta' - d\gamma}{\sqrt{2d\sigma^2}} \right)$$

Hence, the proof of inequality (5.5) is complete.

Assume that $S(T_i + d) > s_0 - \delta$. Since the value of the stored energy increases by $S_\mu(T_i) - s_0 + \delta'$ units after $t = T_i + d$, we obtain the following relations:

$$\lim_{\varepsilon \to 0^+} S(T_i + d + \varepsilon) = S_\mu(T_i) - s_0 + \delta' + S(T_i + d)$$
$$\geq S_\mu(T_i) + \delta' - \delta \geq S_\mu(T_i) - \delta' \geq s_0 - \delta' \quad \blacksquare$$

Concerning Theorem 5.1, the lower bound of our system reliability ratio is obtained in the following way if each community gets help from the auxiliary power plants (in the way mentioned in algorithm 5.1):

$$\mathbf{Pr}[\forall t \in [T_i, T_{i+1}] : S(t) > s_0 - \delta] = \mathbf{Pr}[\forall t \in [T_i, T_i + d] : S(t) > s_0 - \delta]$$
$$\times \mathbf{Pr}[\forall t \in (T_i + d, T_{i+1}] : S(t) > s_0 - \delta | \forall t \in [T_i, T_i + d] : S(t) > s_0 - \delta]$$

$$\to \mathbf{Pr}[\forall t \in [T_i, T_{i+1}] : S(t) > s_0 - \delta] \geq \mathrm{erf}\left(\frac{\delta - \delta' - d\gamma}{\sqrt{2d\sigma^2}}\right) \times 1$$

$$\to \mathbf{Pr}[\forall t \in [T_i, T_{i+1}] : S(t) > s_0 - \delta] \geq \mathrm{erf}\left(\frac{\delta - \delta' - d\gamma}{\sqrt{2d\sigma^2}}\right)$$

As a numerical example of the obtained lower bound of the reliability ratio in time interval $[T_i, T_{i+1}]$, consider the following values for the contributing parameters: $\delta = 2\delta' = 102\,\mathrm{kWh}$, $d = 1/5\,\mathrm{h}$, $\sigma^2 = 40\,\mathrm{kW^2h}$, and $\gamma = 185\,\mathrm{kW}$. In this case, the lower bound of the reliability ratio in $[T_i, T_{i+1}]$ is approximately $1 - 7 \times 10^{-7}$.

The important question that has not yet been answered in our model is how much energy the auxiliary power plants have to generate to satisfy community demand. In this section, we find a low threshold of the amount of power generated by the auxiliary power plants, which guarantees that the community demand is satisfied with high probability. We first find a lower bound for probability $\mathbf{Pr}[T_{i+1} - T_i > t]$:

Theorem 5.2 Assuming that $T_{i+1} < t_{\mathrm{end}}$, if the amount of stored energy does not meet low threshold $s_0 - \delta$ in the time interval $[T_i, T_i + d]$, the following inequality holds:

$$\mathbf{Pr}[T_{i+1} - T_i > t] > \mathrm{erf}\left(\frac{\delta'}{\sqrt{2t\sigma^2}}\right) \tag{5.7}$$

Proof. For any $x \in [d, T_{i+1} - T_i]$, we obtain the following equations:

$$S(T_i + x) = S_{\mathrm{aux}}(T_i + x) + \int_0^{T_i + x} (\mathrm{Gen}(t') - \mathrm{Dem}(t'))\mathrm{d}t'$$
$$S(T_i) = S_{\mathrm{aux}}(T_i) + \int_0^{T_i} (\mathrm{Gen}(t') - \mathrm{Dem}(t'))\mathrm{d}t'$$

Considering the last two equations, by subtracting the latter equation from the earlier one, we obtain the following relations:

$$\mathcal{S}(T_i + x) - \mathcal{S}(T_i) = \underbrace{\mathcal{S}_{\text{aux}}(T_i + x) - \mathcal{S}_{\text{aux}}(T_i)}_{\text{auxiliary energy in } [T_i, T_i + d]} + \int_{T_i}^{T_i + x} (\text{Gen}(t') - \text{Dem}(t')) dt'$$

$$= \mathcal{S}_\mu(T_i) - s_0 + \delta' + \int_{T_i}^{T_i + x} (\text{Gen}_\mu(t') - \text{Dem}_\mu(t')) dt'$$

$$- \int_{T_i}^{T_i + x} (D_{t'} - G_{t'}) dt'$$

$$= \mathcal{S}_\mu(T_i) - s_0 + \delta' + \mathcal{S}_\mu(T_i + x) - \mathcal{S}_\mu(T_i) - \mathcal{W}_x^{(i)}$$

where $\mathcal{W}_x^{(i)} = \int_{T_i}^{T_i + x} (D_{t'} - G_{t'}) dt'$ is a scaled Wiener process of the following covariance function:

$$\mathbf{cov}(s, t) = \min\{s, t\} \cdot \sigma^2$$

Since $\mathcal{S}(T_i)$ is equal to $s_0 - \delta'$, we obtain the following equation:

$$\mathcal{S}(T_i + x) = \mathcal{S}_\mu(T_i + x) - \mathcal{W}_x^{(i)} \tag{5.8}$$

which implies that:

$$\mathcal{S}(T_i + x) \geq s_0 - \mathcal{W}_x^{(i)} \tag{5.9}$$

Since $\mathcal{S}(t) > s_0 - \delta$ for every $t \in [T_i, T_i + d]$, according to Theorem 3.1, $\mathcal{S}(t') > s_0 - \delta'$ for $t' \to (T_i + d)^+$. Subsequently, algorithm 5.1 continues after time $T_i + d$ to find the value of T_{i+1}. In the case that $\mathcal{S}(T_i + x) > s_0 - \delta'$ for every $x \in (d, t]$, the algorithm cannot find T_{i+1}, i.e., $T_{i+1} - T_i > t$. Henceforth:

$$\mathbf{Pr}[T_{i+1} - T_i > t] \geq \mathbf{Pr}[\forall x \in (d, t] : \mathcal{S}(T_i + x) > s_0 - \delta']$$

According to inequality (5.9), we conclude that:

$$\mathbf{Pr}[T_{i+1} - T_i > t] \geq \mathbf{Pr}[\forall x \in (d, t] : \mathcal{W}_x^{(i)} < \delta']$$

$$\geq \mathbf{Pr}[\sup_{d < x \leq t} \mathcal{W}_x^{(i)} < \delta']$$

$$\geq \mathbf{Pr}[\sup_{0 \leq x \leq t} \mathcal{W}_x^{(i)} < \delta']$$

$$\geq \mathbf{Pr}[M_t^{(i)} < \delta']$$

where $M_t^{(i)}$ is the running maximum process corresponding to $\mathcal{W}_x^{(i)}$. So, we obtain inequality (5.7). ∎

Regarding algorithm 5.1, the total amount of electric energy that the auxiliary power plants flow into a community in time interval $[T_i, T_{i+1}]$ is equal to $\mathcal{S}_\mu(T_i) - s_0 + \delta'$. Let $\bar{P}_i$ denote the average power demand of each community in this interval (for $i = 2, 3, \ldots$). Here, we find a lower bound for the value of $\bar{P}_i$.

$$\bar{P}_i = \frac{\text{energy request in } [T_i, T_{i+1}]}{T_{i+1} - T_i}$$

$$= \frac{S_\mu(T_i) - s_0 + \delta'}{T_{i+1} - T_i}$$

$$\leq \frac{S_{\max} - s_0 + \delta'}{T_{i+1} - T_i}$$

Additionally, concerning inequality (5.7), the length of interval $[T_i, T_{i+1}]$ does not exceed t with the probability more than $\text{erf}\left(\delta'/\sqrt{2t\sigma^2}\right)$. Subsequently, we obtain the following inequality regarding the average power demand of each community.

$$\mathbf{Pr}\left[\bar{P}_i < \frac{S_{\max} - s_0 + \delta'}{t}\right] > \text{erf}\left(\frac{\delta'}{\sqrt{2t\sigma^2}}\right) \quad \forall i = 2, 3, \dots$$

Let m and n denote the number of communities and auxiliary power plants, respectively. Moreover, assume that $T_i^{(j)}$, δ_j', and σ_j^2 denote the values of T_i, δ', and σ^2, respectively, for the jth community. Theorem 5.2 implies that:

$$\mathbf{Pr}\left[T_{i+1}^{(j)} - T_i^{(j)} > t\right] > \text{erf}\left(\frac{\delta_j'}{\sqrt{2t\sigma_j^2}}\right) \quad \forall j = 1, 2, \dots, m$$

Or equivalently,

$$\mathbf{Pr}\left[T_{i+1}^{(j)} - T_i^{(j)} \leq t\right] < 1 - \text{erf}\left(\frac{\delta_j'}{\sqrt{2t\sigma_j^2}}\right) \quad \forall j = 1, 2, \dots, m$$

which implies that:

$$\mathbf{Pr}\left[\forall j = 1, 2, \dots, m: T_{i+1}^{(j)} - T_i^{(j)} > t\right] = 1 - \mathbf{Pr}\left[\exists j = 1, 2, \dots, m: T_{i+1}^{(j)} - T_i^{(j)} \leq t\right]$$

$$= 1 - \mathbf{Pr}\left[\bigvee_{j=1}^m T_{i+1}^{(j)} - T_i^{(j)} \leq t\right]$$

$$\geq 1 - \sum_{j=1}^m \mathbf{Pr}\left[T_{i+1}^{(j)} - T_i^{(j)} \leq t\right]$$

Subsequently, we obtain the following inequality:

$$\mathbf{Pr}\left[\forall j = 1, 2, \dots, m: T_{i+1}^{(j)} - T_i^{(j)} > t\right] \geq 1 - m + \sum_{j=1}^m \text{erf}\left(\frac{\delta_j'}{\sqrt{2t\sigma_j^2}}\right) \quad (5.10)$$

In a similar way, we obtain another inequality regarding the average power demand of the jth community in interval $\left[T_i^{(j)}, T_{i+1}^{(j)}\right]$ (expressions $S_{\max}^{(j)}$ and $s_0^{(j)}$ represent the values of $S_{\max}$ and s_0 for the jth community, respectively).

$$\mathbf{Pr}\left[\forall j = 1, 2, \dots, m: \bar{P}_i^{(j)} < \frac{S_{\max}^{(j)} - s_0^{(j)} + \delta_j'}{t}\right] \geq 1 - m + \sum_{j=1}^m \text{erf}\left(\frac{\delta_j'}{\sqrt{2t\sigma_j^2}}\right) \quad (5.11)$$

where $\bar{P}_i^{(j)}$ denotes the average amount of the power demand of the jth community in time interval $\left[T_i^{(j)}, T_{i+1}^{(j)}\right]$.

Concerning inequality (5.11), we can find a probabilistic upper bound over the average total throughput $\bar{P}$ of the n APPs:

$$\mathbf{Pr}\left[\bar{P} < \sum_{j=1}^{m} \frac{S_{\max}^{(j)} - s_0^{(j)} + \delta_j'}{t}\right] \geq 1 - m + \sum_{j=1}^{m} \mathrm{erf}\left(\frac{\delta_j'}{\sqrt{2t\sigma_j^2}}\right)$$

which implies that with high probability, we can bound the total power demand of all the communities. Consequently, the assumption that we have few APPs in the city (in comparison with m) is reasonable.

As a numerical example, let $m = 200$, $n = 4$. In addition, consider that for every community, $S_{\max} = 180$ kWh, $s_0 = 140$ kWh, $\delta' = 51$ kWh, and $\sigma^2 = 40$ kW^{2}h. Assuming that $p(t)$ denotes the lower bound obtained by inequality (5.10),

$$p(t) = 200 \times \mathrm{erf}\left(\frac{51}{\sqrt{80t}}\right) - 199$$

By setting $t = 3.61$ h, we obtain $p(t) \simeq .9956$. Finally, we conclude the following statement regarding our electrical system example:

With a certainty of more than 99.56%, the duration between any two successive energy requests of every community is more than 3.61 h.

Since after each request, the auxiliary power plants send a flow of at most $S_{\max} - s_0 + \delta' = 91$ kWh energy into the target community, the total power demand of the communities does not exceed $\dfrac{91 \times 200}{3.61} \simeq 5$ MW with a certainty of more than 99.56%.

5.3 Analysis of the Energy Distribution Cost

In section 5.1, we described an example of the residential electricity system spread in a city. Additionally, the network representation of this system and how the electric energy flows inside each community was addressed. In the second section, further information about the energy flows into the community was specified. Furthermore, we discussed how auxiliary power plants can improve the reliability of our electric system.

Although the auxiliary power plants make our system more reliable, flowing energy into the community does incur some cost. Our goal in this section is to minimize the energy flow cost. At first, we introduce the system global controller, which manages the energy requests of the local controllers in the FCFS[3] order. Then, we discuss the energy flow cost function.

5.3.1 Global Controller

As mentioned in section 5.2, in each community, there is a local controller running algorithm 5.1. In line 8 of this algorithm, the controller requests $E_i = (S_\mu(T_i) - s_0 + \delta')$ units of energy. There exists a global controller in the city to manage the energy requests made by different communities.

In the rest of this section, let sequence C_1, C_2, ..., C_m denote the m communities located in the city. Also, assume that $E_i^{(j)}$ represents the amount of energy requested by community C_j in time $t = T_i^{(j)}$, where $T_i^{(j)}$ is defined as before. Each community expects that its request will be satisfied at most d hours after making it. Every energy request from community C_j is represented as a tuple of two values: the index of the target community j and the amount of requested energy $E_i^{(j)}$. When a community requests some amount of energy, its local controller sends the associated tuple of the request to the global controller. Then, the global controller adds the tuple to the queue *reqQueue* where requests are waiting for response. All of the queued requests will be processed after some waiting time. Algorithm 5.2 and algorithm 5.3 are two concurrently running threads that specify the functionality of the global controller in detail.

As algorithm 5.2 shows, the global controller is always waiting to receive a new energy request and add it into the queue of requests. One the other hand, algorithm 5.3, which is running in parallel with the earlier algorithm, does nothing unless function Now() returns a multiple of $d/2$ and the queue of requests is not empty. In this case, the controller dequeues all of the queued requests and then initiates their responding process (lines 7

	Algorithm 5.2 Global controller (thread 1)
1	**while true do**
2	**if** there is a new request **then**
3	l reqQueue.Enqueue(GetRequest());
4	**end**
5	**end**

---

3. First come first served.

Algorithm 5.3   Global controller (thread 2)

---

1 **while true do**
2 $\quad$ **if** Now() = multiple of $\frac{d}{2}$ **then** //d is the maximum response time of each request
3 $\qquad$ **if** reqQueue.Size() = 0 **then**
4 $\qquad\quad$ **continue**;
5 $\qquad$ **end**
6 $\qquad$ reqQueue.Dequeue(reqList,reqQueue.Size()); //reqList is a list of request tuples. Function Dequeue(list,n) dequeues n request and puts it in list.
7 $\qquad$ genStatus $\leftarrow$ *GetGeneratorsStatus*();
8 $\qquad$ commdityList $\leftarrow$ Route(reqList, genStatus);
9 $\qquad$ StartFlow(commodityList);
10 $\quad$ **end**
11 **end**

---

to 9). The initiated process, which takes at most $d/2$ hours, is done in three consequent steps: 1. Ask auxiliary power plants (APPs) how much energy they can output (maximum throughput) in the upcoming $d/2$ hours (line 7); 2. Specify all of the energy flows that are used to satisfy the energy demand of the target communities (line 8); and 3. Start flowing energy into the target communities (line 9). Here, we address these three steps in detail.

Step 1 Let sequence $A_1, A_2, ..., A_n$ denote the n APPs in the city. At the first step, the global controller asks the APPs about the maximum amount of their energy output in the upcoming $d/2$ hours. Assume that $G_i(k)$ represents the maximum energy amount that APP A_i can produce over the time interval $[kd/2, kd/2 + d/2)$. Subsequently, function *GetGeneratorsStatus*() in algorithm 5.3 returns the following information about the APPs in $t = kd/2$:

$$(1, G_1(k)), (2, G_2(k)), ..., (n, G_n(k))$$

such that the ith ordered pair specifies the maximum throughput of APP A_i over time interval $[kd/2, kd/2 + d/2)$.

Step 2 Assume that every energy request is represented by ordered pair (v, X) such that v is the index of the target community and X is the amount of requested energy. Additionally, let the following sequence denote the queued energy requests at time $t = kd/2$:

$$(n_1(k), \mathcal{E}_1(k)), (n_2(k), \mathcal{E}_2(k)), \dots, (n_q(k), \mathcal{E}_q(k))$$

Also, consider $G = (V,E)$ as the graph representation of our grid such that:

$$V = \{C_i : i = 1, 2, \dots, m\} \cup \{A_i : i = 1, 2, \dots, n\}$$

Set V is the set of nodes (either communities or APPs) and E is the set of edges representing the two-way transmission power lines connecting the nodes. In this step, the global controller has to compute the following energy flows derived from the n APPs toward every target community $C_{n_j(k)}$:

$$\left(p_{1j}^{(k)}, x_{1j}^{(k)}\right), \left(p_{2j}^{(k)}, x_{2j}^{(k)}\right), \dots, \left(p_{nj}^{(k)}, x_{nj}^{(k)}\right)$$

where $p_{ij}^{(k)}$ is a path from A_i to $C_{n_j(k)}$ in graph G (for every $i = 1, 2, \dots, n$, $k \in \mathbb{N}$, and $j = 1, 2, \dots, q$), and $x_{ij}^{(k)}$ values are restricted by the following conditions:

$$\begin{cases} x_{ij}^{(k)} \geq 0 & \forall j = 1, 2, \dots, q, \quad \forall i = 1, 2, \dots, n \\ \sum_{i=1}^{n} x_{ij}^{(k)} = \mathcal{E}_j(k) & \forall j = 1, 2, \dots, q \quad \forall k \in \mathbb{N} \\ \sum_{j=1}^{q} x_{ij}^{(k)} = G_i(k) & \forall i = 1, 2, \dots, n \end{cases} \tag{5.12}$$

Before describing the third step, we use inequality (5.7) to find a probabilistic high threshold for the number of queued requests (q). Assume $\gamma d/2$ as an upper bound for the average of random variable $(T_{i+1} - T_i)$ for some γ. By letting $t = \gamma d/2$ in inequality (5.7), we obtain the following result:

$$\Pr[T_{i+1} - T_i > t] > \text{erf}\left(\frac{\delta'}{\sqrt{d\gamma\sigma^2}}\right)$$

Additionally, for some integer $r > 0$, T_r can be written as the summation of r i.i.d. variables:

$$T_r = \sum_{i=0}^{r-1}(T_{i+1} - T_i)$$

Considering the Central Limit Theorem, if $r \to +\infty$, variable T_r converges in distribution to $\mathcal{N}(r\gamma d/2, r\sigma_T^2)$, where σ_T^2 denotes the variance of random variable $(T_{i+1} - T_i)$.

Now, let Q denote the Binomial random variable representing the upper bound of the number of queued requests in period $[kd/2, kd/2 + d/2)$. The success probability of Q is equal to the maximum of the following expression:

$$\Pr\left[T_r \in \left[\frac{kd}{2}, \frac{kd}{2} + \frac{d}{2}\right]\right] \leq 2\text{erf}\left(\frac{d}{4\sqrt{2r\sigma_T^2}}\right) = p$$

Henceforth, regarding Hoeffding's Inequality, we obtain the following:

$$\Pr[Q > q'] = 1 - F_Q(q') \le e^{-2\epsilon k} \qquad (5.13)$$

where $\epsilon = q'/k - p$. Note that we can estimate σ_T^2 using the variance of the running maximum process mentioned in the previous section:

$$\sigma_T^2 \simeq \left(1 - \frac{2}{\pi}\right)\sigma^2 \frac{d}{2}$$

which implies that:

$$p \simeq 2\mathrm{erf}\left(\frac{1}{4}\sqrt{\frac{d\pi}{r\sigma^2(\pi-2)}}\right)$$

Note that in the above analysis, for simplicity, we assumed that for every community C_j, the variance of the Wiener process representing its storage energy is identical and equal to σ^2.

Note that we can bound the maximum number of queued requests (q) with high probability using inequality (5.13).

Step 3   All of the energy flows specified in the previous step are triggered in this step at the same time. Assume that for every flow $f = (p, x)$ obtained in the second step, the flow continues for T hours where T is a constant value *independent* of p and x. Consequently, it takes T hours to finish step 3. If t' and t'' represent the duration of steps 1 and 2, respectively, then we assume that the value of $t' + t''$ is too small in comparison with T to be taken into account. Considering $T = d/2$, we conclude that it takes $d/2$ hours to respond the queued energy requests. In addition, regarding algorithm 5.3, the maximum waiting time of each request in queue *reqQueue* does not exceed $d/2$. Hence, the total time it takes to satisfy a community's energy request is at most $d/2 + d/2 = d$ hours.

5.3.2   Energy Flow Cost Function

As mentioned before, in our residential electric system, the demand of every consumer in the city is satisfied by two energy resources: the renewable power plant located inside the community in which the consumer resides, and the APPs scattered across the city. In the case where the production of a renewable power plant does not satisfy consumer demand completely, the APPs flow electricity through a path toward the *target community* in which the consumer is located. Each electricity flow path may contain a number of communities and APPs (see figure 5.6). All of the flows are triggered by the global controller every $d/2$ hours (line 9 of algorithm 5.3). Each flow

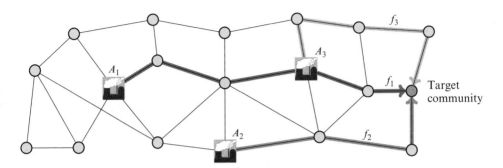

Figure 5.6 Electricity flow into the shown target community. Note that there are three energy flows derived from different APPs.

has a specific magnitude (which is measured as the amount of transmitted energy per time unit); however, all of the flows last the same amount of time $(d/2)$. Note that some of the APPs may flow no energy into a target community; in such a case, the magnitude of their associated flows is assumed to be zero. Additionally, in any period of $d/2$ hours, there may be multiple target communities in the city.

Edge Cost Function Consider figure 5.6 again. Transmitting energy between any two adjacent nodes in the network of communities and APPs incurs some cost. We assume that this cost is divided into two parts:

1. The cost of energy loss in cables connecting two nodes.
2. The cost incurred in the nodes.

The both cost portions are functions of the transmission power. As a result, we obtain the following equation:

$$\text{EdgeCost}(p) = l(p) + n(p) \quad \forall \text{ transmission power } p$$

where $l(p)$ and $n(p)$ represent the cost of energy loss and the cost incurred at the endpoints, respectively. For small values of p, the cost $l(p)$ is too small, in comparison with $n(p)$, to be taken into account. However, since $l(p)$ is an increasing, convex function, by increasing the transmission power, the value of $l(p)$ becomes greater, and dl/dp gradually increases. Furthermore, the cost incurred at the endpoints of a power transmission is assumed to be an increasing concave function. Note that the initialization of an energy transmission session incurs a constant amount of cost at the endpoints regardless of how much power is transmitted. Consequently, as the transmission power increases, the value of dl/dp becomes lower, i.e., function $n(p)$ is concave. Figure 5.7 illustrates both the cost portions, $l(p)$ and $n(p)$,

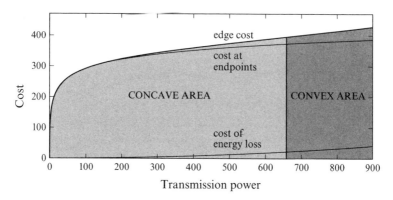

Figure 5.7 Plot of the edge cost function (the black plot), which is the summation of the energy loss cost (lower plot) and the cost incurred at the endpoints (the middle plot).

.

and the edge cost itself (which is sum of the portions) as functions of transmission power.

As figure 5.7 shows, the edge cost plot is divided into two domains. At the first domain (*concave area*), the cost $l(p)$ is so small that the edge cost is almost equal to $n(p)$. Consequently, EdgeCost(p) is a concave function. On the other hand, the edge cost in the second domain (*convex area*) becomes a convex function of transmission power. In other words, the increase rate of dl/dp becomes greater than the decrease rate of dl/dp.

We will assume for the rest of this section that every energy transmission between any two nodes in our example occurs in the concave area, i.e., the transmission power does not exceed some threshold (i.e., as the amount of energy requests of communities is bounded, the assumption is feasible). Consequently, the transmission cost of power line $e \in E$ is equal to $w_e \cdot \text{rrc}(p)$ where p is the amount of power transmitted through e, rrc is a concave function (relative routing cost function) which is identical for every power line in the grid, and w_e is called the weight of power line e which is considered as a characteristic of e.

5.4   Cost Optimization of Energy Distribution

In this section, we use one of the versatile routing schemes mentioned in chapter 3 to compute the flow paths in our electric system. We also get help from a global optimization method made for minimizing the class of linearly constrained concave functions to compute the flow values through each computed energy path. Finally, we will show that the total cost of the energy distribution will not be more than $O(\log^2 n)$ times the minimum possible cost, where n is the number of nodes (either communities or power plants) in the energy grid.

Note from line 8 of algorithm 5.3 that the global controller uses the list of the energy requests and the instantaneous status of the APPs to compute the flow paths and the amount of flows toward the target communities (the second step in the aforementioned responding process). Here, we focus on how this step is done by the global controller.

Regarding figure 5.8, we need to compute the energy path $p_{ij}^{(k)}$ for the ith source and the jth target. Additionally, we have to specify how much energy should flow between A_i and $C_{n_j(k)}$ for every $i = 1, 2, \ldots, n$, $j = 1, 2, \ldots, q$, and positive integer k.

Assuming that $\mathbb{S}$ denotes an integral versatile routing scheme in the energy grid, path $p_{ij}^{(k)}$ is computed by the global controller using the following equation:

$$p_{ij}^{(k)} = \mathbb{S}\left(A_i, C_{n_j(k)}\right) \quad \forall i = 1, 2, \ldots, n, \forall j = 1, 2, \ldots, q, \forall k \in \mathbb{N}$$

Furthermore, the global controller determines the energy flow value $x_{ij}^{(k)}$ for every computed path $p_{ij}^{(k)}$. The value of every flow is computed in a way that the total incurred cost in time interval $[kd/2, kd/2 + d/2)$ gets its *minimum possible* value (concerning the paths suggested by scheme $\mathbb{S}$). In the kth period, the energy distribution problem can be mapped into a general routing problem with the following total flow cost:

$$\text{Total cost} = \sum_{e \in E} w_e \cdot \text{rrc}(f_e(k))$$

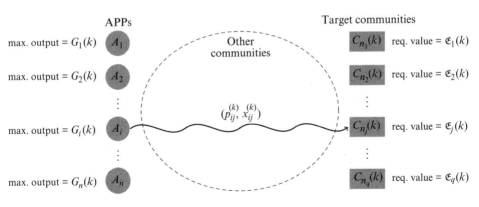

Figure 5.8 Schematic view of the energy distribution through the city over time interval $[kd/2, kd/2 + d/2)$.

where $f_e(k)$ represents the amount of flow passing through edge e in time interval $[kd/2, kd/2 + d/2)$:

$$f_e(k) = \sum_{i=1}^{n} \sum_{j=1}^{q} y_{ij}^{(k)}(e) \quad \forall e \in E, \forall k \in \mathbb{N}$$

and

$$y_{ij}^{(k)}(e) = \begin{cases} x_{ij}^{(k)} & e \in p_{ij}^{(k)} \\ 0 & \text{otherwise} \end{cases} \quad \forall i = 1, 2, \ldots, n, \forall j = 1, 2, \ldots, q, \forall k \in \mathbb{N}, \forall e \in E$$

Henceforth, every flow value $x_{ij}^{(k)}$ will be an optimal solution of the following linearly constrained concave optimization problem:

$$\text{minimize} \sum_{e \in E} w_e \cdot \text{rrc}(f_e(k)) \tag{5.14}$$

subject to condition (5.12).

To minimize the mentioned linearly restricted concave function, we use a method presented by J. B. Rosen that computes the global solution by partition of feasible domain. This method will be briefly described in the next section.

5.4.1 Computation of Energy Flow Values Using Rosen's Method

Consider the amount $x_{ij}^{(k)}$ of energy that flows from APP A_i to community $C_{n_j}(k)$ through path $p_{ij}^{(k)}$ for every $i = 1, 2, \ldots, n$ and $j = 1, 2, \ldots, q$ in the kth period of time. Hence, for a specific k, we need to solve the minimization problem, which is specified in equation (5.14) and has $n \cdot q$ variables to determine all of the flow values. Additionally, since these variables are restricted by condition (5.12), it is the case that:

$$x_{nj}^{(k)} = \mathcal{E}_j(k) - \sum_{i=1}^{n-1} x_{ij}^{(k)} \quad \forall j = 1, 2, \ldots, q$$

which means that we can reduce the number of variables to $(n-1)q$ with the following restrictions:

$$\begin{cases} -x_{ij}^{(k)} \leq 0 & \forall j = 1, 2, \ldots, q, i = 1, 2, \ldots, n-1 \\ \sum_{i=1}^{n-1} x_{ij}^{(k)} \leq \mathcal{E}_j(k) & \forall j = 1, 2, \ldots, q \\ \sum_{j=1}^{q} x_{ij}^{(k)} \leq G_i(k) & i = 1, 2, \ldots, n-1 \\ -\sum_{j=1}^{q} \sum_{i=1}^{n-1} x_{ij}^{(k)} \leq G_n(k) - \sum_{j=1}^{q} \mathcal{E}_j(k) \end{cases}$$

Subsequently, we can formulate the minimization problem in the following form:

$$\begin{aligned} \text{minimize} \quad & \phi(x) \\ \text{subject to} \quad & x \in \Omega \subseteq \mathbb{R}^{nq-q} \end{aligned} \tag{5.15}$$

such that function $\phi(x)$ is concave and twice differentiable and Ω denotes a bounded area (polyhedron) which is specified in the following way:

$$\Omega = \{x : Ax + x = b \wedge z \geq 0\} \tag{5.16}$$

where the constraint coefficient matrix A is $(nq + n) \times (nq - q)$ and *slack vector* z belongs to $\mathbb{R}^{nq+q}$.

In 1982, J. B. Rosen presented a computationally difficult method to solve this global minimization problem. Here, we briefly describe the method. Note that the global solution of the problem is denoted by $v^* \in \Omega$ and also ϕ^* represents the minimum value of $\phi(x)$ in polyhedron Ω, i.e.:

$$\phi^* = \phi(v^*)$$

Additionally, we assume that $a = nq - q$ denotes the number of variables and $b = nq + n$ represents the number of constraints on the variables.

Initially, we find a vertex $\hat{v}$ of polyhedron Ω such that $\hat{\phi} = \phi(\hat{v})$ gives a reasonable upper bound for the minimum value ϕ^*. Let $u_1, u_2, ..., u_a$ denote the *eigenvector*[4] of function $\phi(x)$*Hessian*[5] matrix such that $\|u_i\| = 1$ for every $i = 1, 2, ..., a$. Let $\bar{x}$ denote a computed local maximum of $\phi(x)$ in Ω. Since ϕ is a concave function, point $\bar{x}$ also specifies a global maximum for $\phi(x)$ in Ω. Now, we can estimate $\phi(x)$ using the following quadratic function:

$$\phi(x) \simeq \phi(\bar{x}) + \nabla\phi^T(\bar{x})(x-\bar{x}) + \frac{1}{2}(x-\bar{x})^T H_\phi(\bar{x})(x-\bar{x})$$

where $\nabla\phi$ and H_ϕ denote the *gradient* and the Hessian matrix of function ϕ, respectively. Using the intuition that the aforementioned vertex v^* should be farthest from the global maximum point $\bar{x}$ in all the directions $\pm u_i$ (for every $i = 1, 2, ..., n$), we obtain the following equation:

$$\hat{\phi} = \min_{i=1}^{2a} \phi(v_i) \tag{5.17}$$

such that:

$$v_i = \max_{x \in \Omega} w_i^T x \quad \forall i = 1, 2, ..., 2a \tag{5.18}$$

and

4. An eigenvector of a square matrix A is a nonzero vector v such $Av = \lambda v$ for some constant λ.
5. Hessian matrix $A = [a_{ij}]_{n \times n}$ of twice differentiable function $f(x_1, x_2, ..., x_n)$ is a square matrix of second-order partial derivatives of f such that for every $i, j = 1, 2, ... , n$,

$$a_{ij} = \frac{\partial^2 f}{\partial x_i \partial x_j} .$$

$$w_i = -w_{a+i} = u_i \quad \forall i = 1, 2, \ldots, a$$

Equation (5.18) denotes the set of $2a$ linear optimization problems that can be solved using normal LP[6] calculations. In solving these problems, we use equation (5.17) to find vertex $\hat{v}$, which makes the value of function ϕ minimum (among all of the $2a$ computed vertices):

$$\hat{\phi} = \phi(\hat{v})$$

After computing $\hat{\phi}$ as an upper bound of ϕ^*, we construct two rectangular domains: $\bar{R} \supseteq \Omega$ and $R(\hat{\phi})$. Domain $\bar{R}$ is obtained by intersecting the following $2a$ half-spaces:

$$\bar{R} = \cap_{i=1}^{2a} \bar{H}_i$$

such that:

$$\bar{H}_i = \left\{ x : w_i^T (x - \bar{x}) \le \beta_i \right\} \quad \forall i = 1, 2, \ldots, 2a$$

and

$$\beta_i = w_i^T (v_i - \bar{x})$$

Concerning equation (5.18), it is easy to see that $\Omega \subseteq \bar{H}_i$ for every $i = 1, 2, \ldots, 2a$. Since $\Omega \subseteq \bar{R}$, we will obtain a lower bound for ϕ^* by finding the minimum of function ϕ over polyhedron $\bar{R}$. Since ϕ is a concave function, it gets its minimum over $\bar{R}$ at one of the 2^a vertices of $\bar{R}$. Assuming $\underline{\phi}$ as this minimum value, we deduce that:

$$\underline{\phi} = \phi(\bar{x}) + \sum_{i=1}^{a} \min \{\alpha_i, \alpha_{n+i}\}$$

where:

$$\alpha_i = \beta_i \nabla \phi^T (\bar{x}) w_i - \frac{1}{2} \beta_i^2 \bar{\lambda}_i \quad \forall i = 1, 2, \ldots, 2a$$

and considering $\lambda_1, \lambda_2, \ldots, \lambda_a$ as the sequence of eigenvalues of Hessian matrix of function $\phi(x)$, $\bar{\lambda}_i$ is defined in the following way:

$$\bar{\lambda}_i = \bar{\lambda}_{i+a} = \lambda_i \quad \forall i = 1, 2, \ldots, a$$

We have already computed a lower bound ($\underline{\phi}$) and upper bound ($\hat{\phi}$) for the value of ϕ^*. In the case that $\varepsilon = \hat{\phi} - \underline{\phi}$ is small enough, we can claim that $\hat{v}$ attains a global minimum value of function ϕ over domain Ω with an acceptable precision and end the calculation.

Additionally, we construct domain $R(\hat{\phi})$ which is defined as:

$$\left(x \in R(\hat{\phi}) \right) \leftrightarrow \left(\phi(x) \ge \hat{\phi} \right)$$

6. Linear program.

Regarding the definition of $R(\hat{\phi})$, it is inscribed in the *ellipsoid*[7] $\phi(x) = \hat{\phi}$ at each of its 2^a vertices. The faces of $R(\hat{\phi})$ are determined by $2a$ hyperplanes in the following forms:

$$H_i = \{x : w_i^T (x - \bar{x}) = \gamma_i\} \tag{5.19}$$

such that:

$$\gamma_i \bar{\lambda}_i = \nabla \phi^T (\bar{x}) w_i + \left(\left(\nabla \phi^T (\bar{x}) w_i \right)^2 + 2\bar{\lambda}_i \left(\frac{\phi(\bar{x}) - \hat{\phi}}{a} \right) \right)^{\frac{1}{2}} \quad \forall i = 1, 2, \dots, 2a$$

Since we have already computed $\hat{v}$ as a feasible point, the global minimum of $\phi(x)$ over Ω does note exceed $\hat{\phi}$. This implies that it is safe to eliminate the interior of domain $R(\hat{\phi})$ from further consideration. Additionally, we use the $2a$ hyperplanes defined in equation (5.19) to partition feasible domain Ω to at most $2a$ convex domains specified in the following equation:

$$\Omega_i = H_i^+ \cap \Omega \quad \forall i = 1, 2, \dots, 2a$$

where H_i^+ denotes the half-space corresponding to H_i and exterior to domain $R(\hat{\phi})$. Concerning equation (5.18), it is easy to see that:

$$(\Omega_i = \varnothing) \leftrightarrow (v_i \in H_i^+)$$

Assume that we only have r nonempty subdomains (say $\Omega_1, \Omega_2, \dots, \Omega_r$) after partitioning $\Omega - R(\hat{\phi})$ where $r = 1, 2, \dots, 2a$. Henceforth, our global minimization problem breaks into the following r subproblems:

$$\min_{x \in \Omega_i} \phi(x) \quad \forall i = 1, 2, \dots, r$$

such that Ω_i is a polyheron containing vertex v_i which has been previously computed in equation (5.18). For every subdomain Ω_i, we construct a close approximation of it in the form of a simplex denoted by S_i such that $\Omega_i \subseteq S_i$. Here is the specification of the $a + 1$ vertices of simplex S_i:

$$v_i, v_{i1}, v_{i2}, \dots, v_{ia}$$

such that:

$$\begin{cases} v_{ij} = v_i - \theta_{ij} \cdot p_{ij} & \forall i, j = 1, 2, \dots, a \\ \theta_{ij} = \dfrac{u_i^T (v_i - \bar{x}) - \gamma_i}{u_i^T p_j} \end{cases}$$

7. Note that since we have estimated $\phi(x)$ using a quadratic function, $\phi(x) = const$ represents an ellipsoid in a dimensions.

and vectors $P_i = [p_{ij}]_{a \times 1}$ ($i = 1, 2, ..., a$) are obtained from the inverse of the basis matrices corresponding to the constraints of the linear optimization problems specified in equation (5.17).

Since the minimum value of function $\phi(x)$ over simplex S_i is obtained in one of its vertices, if for every $j = 1, 2, ..., a$, $\phi(v_i) \leq \phi(v_{ij})$, we can eliminate S_i from further consideration (note that $\phi(v_i) \geq \hat{\phi}$). By solving the following minimization problems for all of the remaining simplexes, we will obtain a number of lower bounds for the value of $\phi(x)$ over subdomain Ω_i:

$$\underline{\phi}^{(i)} = \min_{x \in S_i} \phi(x)$$

Assuming that $\underline{\phi}' = \min_i \underline{\phi}^{(i)}$, the following two cases apply:

1. The value of $\hat{\phi} - \underline{\phi}'$ does not exceed an acceptable threshold (ε): in this case, vertex $\hat{v}$ attains the global minimum over Ω and we are done.
2. $\hat{\phi} > \underline{\phi}' + \varepsilon$: in this case, we treat each of the nonempty subdomains in the same way as domain Ω was treated.

Note that the implementation of this method is computationally hard; however, regarding the size of problem mentioned in equation (5.14), we need to use this method for $a = O(nq)$ such that n is the number of APPs, which was previously assumed to be much smaller than the number of communities; and q which is the number of target communities in a period of $d/2$ units. As mentioned before, the value of q does not exceed some threshold with high probability, see inequality (5.13). Consequently, it is feasible to use the described method, because the size of our problem is not high.

5.4.2 Competitive Ratio of Cost Distribution

Here, we present a theorem that introduces an upper bound for the cost of the energy flow incurred in the city.

Theorem 5.3 Assuming that C and C^* denote the total incurred cost of energy flow in our solution and in the optimal energy cost, the following inequality holds:

$$\frac{C}{C^*} \leq CR(\mathbb{E}, \mathbb{S})$$

such that $\mathbb{S}$ denotes the versatile routing scheme used for computing the energy flow paths and also $\mathbb{E}$ represents the set of all the possible routing cost environments with concave edge routing cost function, $nrc_\pi = \Sigma$, and the sequence of commodities with any non-negative value from any arbitrary APP to any community in the city.

Proof. Since we computed the flow values using a global minimization method, it is easy to see that the total competitive ratio of our solution is equal to the competitive ratio of the routing scheme we have used for finding the flow paths. ∎

5.5 Summary and Outlook

In this chapter, we introduced a stochastic cost-optimization model for distribution of green power resources in the context of the residential electricity system of a city. In our model, there are a number of communities in a city that are connected, and each one has its own green power plant. Furthermore, we showed that if we have only a few APPs scattered across the city, the reliability of our system improves a lot. To do this, we used a stochastic model of supply and demand in our system.

The power demand of each community is managed by its local flow controller; while there is a global flow controller for managing the power trading between communities. By describing the local and global controllers, we proved that our system incurs at most $\log^2 n$ times as the efficient cost where n denotes the number of green power plants.

Exercises

1. How can we apply the bottom-up versatile routing scheme mentioned in chapter 3 for routing the energy flows through the grid described in this chapter?
   a. What is the approximation ratio of the solution made by this routing scheme?
   b. What is the time-complexity of making the solution? Compare it to the complexity of the method mentioned in Section 3.
2. We assumed that the APPs flow energy in constant-length sessions. How does the assumption make the cost-optimization problem simpler?

Suggested Reading

Banos, R., F. Manzano-Agugliaro, F. G. Montoya, C. Gil, A. Alcayde, and J. Gomez. 2011. Optimization methods applied to renewable and sustainable energy: A review. *Renewable and Sustainable Energy Reviews*, 15(4):1753–1766.

Fakcharoenphol, J., S. B. Rao, and K. Talwar. 2003. A tight bound on approximating arbitrary metrics by tree metrics. *Proceedings of the 35th STOC*, 448–455.

Greatbanks, J. A., D. H. Popovic, M. Begovic, A. Pregelj, and T. C. Green. 2003. On optimization for security and reliability of power systems with distributed generation. *Proceedings of the IEEE PowerTech Conference.* doi:10.1109/PTC.2003.1304111

Gupta, A., M. T. Hajiaghayi, and H. Racke. 2006. Oblivious network design. *Proceedings of the Seventeenth Annual ACM-SIAM Symposium on Discrete Algorithms, SODA '06,* 970–979.

Liu, Y., N. U. Hassan, S. Huang, and C. Yuen. 2013. Electricity cost minimization for a residential smart grid with distributed generation and bidirectional power transactions. *Proceedings of the IEEE Power and Energy Society, IEEE Innovative Smart Grid Technologies* (ISGT). doi: 10.1109/ISGT.2013.6497859.

Natural Resources Defense Council. 2007. Dirty coal is hazardous to your health: Moving beyond coal-based energy. Washington, DC.

Neely, M. J., A. Tehrani, and A. Dimakis. 2010. Efficient algorithms for renewable energy allocation to delay tolerant consumers. *Proceedings of the IEEE First International Conference on Smart Grid Communications, SmartGridComm.* doi:10.1109/ SMARTGRID.2010.5621993

Papavasiliou, A., and S. S. Oren. 2010. Supplying renewable energy to deferrable loads: Algorithms and economic analysis. *Proceedings of the IEEE Power and Energy Society General Meeting.* doi:10.1109/PES.2010.5589405

Papavasiliou, A., S. S. Oren, M. Junca, A. G. Dimakis, and T. Dickhoff. 2008. Coupling wind generators with deferrable loads. *CITRIS Big Ideas White Paper Competition, 3rd Place Winner.*

Sioshansi, R., and W. Short. 2009. Evaluating the impacts of real-time pricing on the usage of wind generation. *IEEE Transactions on Power Systems,* 24(2):516–524.

Rosen, J. B. 1983. Global minimization of a linearly constrained concave function by partition of feasible domain. *Mathematics of Operations Research,* 8(2):215–230.

U.S. Department of Energy. 2008. The smart grid: An introduction. Washington, DC: Office of Electricity Delivery & Energy Reliability.

Vittal, V. 2010. The impact of renewable resources on the performance and reliability of the electricity grid. *The Bridge,* 40(1):5–12.

Bibliography

Alon, N., B. Awerbuch, Y. Azar, N. Buchbinder, and J. S. Naor. 2004. A general approach to online network optimization problems. *Proceedings of the 15th SODA*, 942–951.

Andrews, M. 2004. Hardness of buy-at-bulk network design. *Proceedings of the 45th FOCS*, 115–124.

Azar, Y., E. Cohen, A. Fiat, H. Kaplan, and H. Racke. 2003. Optimal oblivious routing in polynomial time. *Proceedings of the 35th STOC*, 383–388.

Banos, R., F. Manzano-Agugliaro, F. G. Montoya, C. Gil, A. Alcayde, and J. Gomez. 2011. Optimization methods applied to renewable and sustainable energy: A review. *Renewable and Sustainable Energy Reviews*, 15(4):1753–1766.

Bartal, Y. 1996. Probabilistic approximations of metric spaces and its algorithmic applications. *Proceedings of the 37th FOCS*, 184–193.

Busch, C., M. Magdon-Ismail, and J. Xi. 2005. Oblivious routing on geometric networks. *Proceedings of the 17th ACM Symposium on Parallelism in Algorithms and Architectures (SPAA)*, 316–324, Las Vegas, Nevada.

Carter, K. M., R.P. Lippmann, and S. W. Boyer. 2010. Temporally oblivious anomaly detection on large networks using functional peers. *Proceedings of the 10th ACM SigComm conference on Internet measurement*, 465–471.

Chan, H., A. Perrig, and D. Song. 2003. Random key predistribution schemes for sensor networks. *Proceedings of the IEEE Symposium on Security and Privacy*, 197.

Chuzhoy, J., A. Gupta, J. S. Naor, and A. Sinha. 2008.On the approximability of some network design problems, *ACM Transactions on Algorithms*, 4(2): 1–17.

Fakcharoenphol, J., S. B. Rao, and K. Talwar. 2003. A tight bound on approximating arbitrary metrics by tree metrics. *Proceedings of the 35th STOC*, 448–455.

Fraigniaud, P. 2007. The inframetric model for the internet. Technical Report.

Garg, N., R. Khandekar, G. Konjevod, R. Ravi, F. S. Salman, and A. Sinha. 2001. On the integrality gap of a natural formulation of the single-sink buy-at-bulk network design formulation, *Proceedings of the 8th IPCO*, 170–184.

Goel, A., and I. Post. 2009. An oblivious $O(1)$-approximation for single source buy-at-bulk. *Proceedings of the 50th FOCS*, 442–450.

Greatbanks, J. A., D. H. PopoviC, M. BegoviC, A. Pregelj, and T. C. Green. 2003. On optimization for security and reliability of power systems with distributed generation. IEEE Bologna PowerTech Conference, June.

Guha, S., A. Meyerson, and K. Munagala. 2000. Hierarchical placement and network design problems. *Proceedings of the 41st FOCS*, 603–612.

Gupta, A., M. T. Hajiaghayi, and H. Racke. 2006. Oblivious network design. *Proceedings of the Seventeenth Annual ACM-SIAM Symposium on Discrete Algorithms, SODA '06*, 970–979.

Gupta, A., A. Kumar, and T. Roughgarden. 2003. Simpler and better approximation algorithms for network design. *Proceedings of the 35th STOC*, 365–372.

Gupta, A., R. Krauthgamer, and J. R. Lee. 2003. Bounded geometries, fractals, and low-distortion embeddings. *Proceedings of the 44th FOCS*, 534–543.

Gupta, A., M. Pal, R. Ravi, and A. Sinha. 2004. Boosted sampling: Approximation algorithms for stochastic optimization problems. *Proceedings of the 36th STOC*, 417–426.

Harrelson, C., K. Hildrum, and S. B. Rao. 2003. Apolynomial-time tree decomposition to minimize congestion. *Proceedings of the 15th SPAA*, 34–43.

Jeong, J., T. T. Kwon, and Y. Choi. 2010. Host-oblivious security for content-based networks. *Proceedings of the 5th International Conference on Future Internet Technologies*, 35–40.

Kleinberg, J., A. Slivkins, and T. Wexler, 2009. Triangulation and embedding using small sets of beacons. *Journal of the ACM*, 56(6): 1–37.

Konjevod, G., A. W. Richa, and D. Xia. 2008. Dynamic routing and location services in metrics of low doubling dimension. *Proceedings of DISC '08*, 379–393.

Kuhn, F., T. Moscibroda, and R. Wattenhofer. 2005. On the locality of bounded growth. *Proceedings of PODC '05*. New York: ACM, 60–68.

Kurosawa, K., W. Kishimoto, and T. Koshiba. 2008. A combinatorial approach to deriving lower bounds for perfectly secure oblivious transfer reductions. *IEEE Transactions on Information Theory*, 54(6):2566–2571.

Liu, Y., N. Ul Hassan, S. Huang, and C. Yuen. 2013. Electricity cost minimization for a residential smart grid with distributed generation and bidirectional power transactions. *Proceedings of the IEEE Power and Energy Society, IEEE Innovative Smart Grid Technologies* (ISGT). doi: 10.1109/ISGT.2013.6497859.

Natural Resources Defense Council. 2007. Dirty coal is hazardous to your health: moving beyond coal-based energy. Washington, DC.

Neely, M. J., A. Tehrani, and A. Dimakis. 2010. Efficient algorithms for renewable energy allocation to delay tolerant consumers. *Proceedings of the IEEE First International Conference on Smart Grid Communications, SmartGridComm.* doi:10.1109/SMARTGRID.2010.5621993

Papavasiliou, A., and S. S. Oren. 2010. Supplying renewable energy to deferrable loads: Algorithms and economic analysis. *Proceedings of the IEEE Power and Energy Society General Meeting.* doi:10.1109/PES.2010.5589405

Papavasiliou, A., S. S. Oren, M. Junca, A. G. Dimakis, and T. Dickhoff. 2008. Coupling wind generators with deferrable loads. *CITRIS Big Ideas White Paper Competition, 3rd Place Winner.*

Rake, H. 2002. Minimizing congestion in general networks. *Proceedings of the 43rd FOCS,* 43–52.

Rosen, J. B. 1983. Global minimization of a linearly constrained concave function by partition of feasible domain. *Mathematics of Operations Research,* 8(2):215–230.

Salman, F. S., J. Cheriyan, R. Ravi, and S. Subramanian. 2000. Approximating the single-sink link-installation problem in network design. *SIAM Journal on Optimization,* 11(3):595–610.

Sioshansi, R., and W. Short. 2009. Evaluating the impacts of real-time pricing on the usage of wind generation. *IEEE Transactions on Power Systems,* 24(2):516–524.

Srinivasagopalan, S. 2011. Oblivious buy-at-bulk network design algorithms. Doctoral dissertation, Louisiana State University, Baton Rouge, LA.

Srinivasagopalan, S., C. Busch, and S. S. Iyengar. 2009. Brief announcement: Universal data aggregation trees for sensor networks in low doubling metrics. *Algorithmic aspects of wireless sensor networks, Proceedings of the 5th International Workshop, ALGOSENSORS.* Berlin: Springer-Verlag, 151–152.

Srinivasagopalan, S., C. Busch, and S. S. Iyengar. 2011. Oblivious buy-at-bulk in planar graphs. *Workshop on Algorithmic Computing, WALCOM,* IIT-Delhi, 18–20 February.

Srinivasagopalan, S., C. Busch, and S. S. Iyengar. 2012. An oblivious spanning tree for single-sink buy-at-bulk in low doubling-dimension graphs. *IEEE Transactions on Computers,* 61(5):700–712.

Talwar, K. 2002. Single-sink buy-at-bulk LP has constant integrality gap. *Proceedings of the 9th IPCO,* 475–486.

U.S. Department of Energy. 2008. The smart grid: An introduction. Washington, DC: Office of Electricity Delivery & Energy Reliability.

Vittal, V. 2010. The impact of renewable resources on the performance and reliability of the electricity grid. *The Bridge,* 40(1):5–12.

Index